DISCORD

DISCORD

THE STORY OF NOISE

MIKE GOLDSMITH

OXFORD
UNIVERSITY PRESS

Great Clarendon Street, Oxford, OX2 6DP,
United Kingdom

Oxford University Press is a department of the University of Oxford.
It furthers the University's objective of excellence in research, scholarship,
and education by publishing worldwide. Oxford is a registered trade mark of
Oxford University Press in the UK and in certain other countries

First Edition published in 2012
Impression: 1

British Library Cataloguing in Publication Data
Data available

Library of Congress Cataloging in Publication Data
Data available

ISBN 978–0–19–960068–7

Printed in Great Britain by
Clays Ltd, St Ives Plc

To the staff of the Acoustics Group,
National Physical Laboratory

ACKNOWLEDGEMENTS

I wish to thank the following for the very useful information they were so generous as to provide:

James P. Allen, Wilbur Professor of Egyptology at Brown University, for information on ancient Egyptian records, and for the translation cited on p. 21;

Dr Richard Barham, National Physical Laboratory, for information on recent environmental noise reduction projects;

Bernard Berry, BEL Acoustics, for material on the history of the National Physical Laboratory;

Alan Bloomfield, for information about EU noise projects;

Professor Philip Dickenson, Massey University, New Zealand, for information about US airbases and their responses to noise problems in the 1940s;

Chris Clark, Head of Digital Research at the British Library, for information about 1940s London soundscapes;

Professor Ed Greitzer, MIT, USA, for information about the Silent Aircraft Initiative;

Chris Hanson-Abbott, founder chairman of Brigade Electronics plc and inventor of the white sound vehicle movement alarm, for material on p. 266;

Mark Hodnett, National Physical Laboratory, for information about the biological effects of ultrasound;

Dr James Hileman, MIT, USA, for information about the Silent Aircraft Initiative;

Paul D. Jepson, BVMS, Ph.D. DipECZM (Wildlife Population Health), MRCVS, European Veterinary Specialist in Wildlife Population

Health, Senior Research Fellow and Graduate Tutor, Institute of Zoology, Zoological Society of London, for information on organ cavitation in marine mammals and for Fig. 41;

Mark Jiggins, Hoare Lea Acoustics, for guidance about wind farms;

Garret Keitzer, for information about early medical accounts of noise;

Lisa Lavia, Managing Director, Noise Abatement Society, for material about John Connell, including Figs 34, 35, and 36 (Courtesy of the Noise Abatement Society, a registered charity in the United Kingdom. Established in 1959, the Society's remit is to help find solutions to noise pollution problems which it does through working with diverse stakeholders to find innovative, practical and holistic solutions. The Society's founder, John Connell OBE, lobbied the Noise Abatement Act through parliament in 1960 making noise a statutory nuisance in the UK for the first time);

Gianluca Memola, National Physical Laboratory, for information and discussion about EU research projects;

Richard Payne, for information about FSN;

Stephen Robinson, National Physical Laboratory, for material and advice concerning underwater noise;

Adam Shaw, National Physical Laboratory, for information about new medical uses of ultrasound.

CONTENTS

LIST OF FIGURES

ABBREVIATIONS

ANASE	Attitudes to Noise from Aviation Sources in England
ARL	Admiralty Research Laboratory
ASA	Acoustical Society of America
ATOC	Acoustical Thermometry of Ocean Climate
CNR	Composite Noise Rating
dB	decibel
dBA	A-weighted decibel
EEC	European Economic Community
END	European Noise Directive
EPA	Environmental Protection Agency
EU	European Union
FAA	Federal Aviation Administration
GES	Good Environmental Status
HIFU	High Intensity Focused Ultrasound
IPPC	Integrated Pollution Prevention and Control
Hz	hertz
kHz	kilohertz
Lden	a weighted 24-hour average of the noise levels during the day, evening, and night
LFA	Low Frequency Activated Sonar
LRAD	Long Range Acoustic Device
MEDUSA	Mob Excess Deterrent Using Silent Audio
MEMS	MicroElectroMechanical Systems
MIT	Massachusetts Institute of Technology
MMO	marine monitoring officer
MRI	Magnetic Resonance Imaging
NAS	Noise Abatement Society

NCA	Noise Control Act
NEPA	National Environmental Policy Act
NNI	Noise and Number Index
NPL	National Physical Laboratory (UK)
NRDC	Natural Resources Defense Council
PNdB	perceived noise level in decibels
PSYOP	Psychological Operations
RNID	Royal National Institute for Deaf People
SOFAR	SOund Fixing And Ranging
SONAR	Sound Navigation and Ranging
WHO	World Health Organization

INTRODUCTION

'Why are you writing a book about the history of *NOISE?*', people asked. 'And what is noise anyway?', they didn't add. But perhaps they should have. Noise is not quite as simple as it sounds, as we shall soon discover. But first: why a noise book? Because noise is more relevant today—and will be more relevant tomorrow—than ever before, thanks in part to the growth of its sources and in part to the more concerted efforts of governments to quash it. Not that quashing is always relevant: noise is by no means always bad.

Probably what most people mean when they use the word is 'unwanted sound'—and indeed this definition of noise has been in common use since the Middle Ages, or perhaps even earlier. But 'unwanted sound' is hardly a definition likely to satisfy a philosopher or a lawyer: is *any* unwanted sound noise? The key in the lock for a burglar? The voice of an enemy? The theme tune of your least favourite soap? 'Unwanted' here surely refers to a sound that is unwanted in itself, rather than in what it signifies. This makes British physicist George William Clarkson Kaye's 1931 definition as 'sound out of place'[1] a better one than 'unwanted sound': a trumpet is just what is wanted in a piece of jazz, but is not so good in a lounge. The rush of water is in place on a riverbank, but not in a cellar. Does definition matter? After all, everyone knows what noises are, even though it may be difficult to find examples that everyone would agree on. But, actually, definition is key to how noise is dealt with: on small scales and large, individuals, households, local authorities, governments, and supra-national organizations all try, with more or less seriousness and success, to reduce the level of 'noise'—and how can they do this if noise is so subjective and indefinable a thing? Isn't trying to ban it a bit like trying to ban bad music or unfunny comedy?

A definition of noise also allows us to determine whether examples of it share any physical characteristics, which may help in its control. And, in addition to giving people a target to agree on, discuss, and battle, the identification of physical indicators is also necessary if noise is to be measured—as is essential for both legal and scientific purposes. Luckily, then, there are types of sound that do share certain physical characteristics and that are widely thought of as being noises. Such sounds tend to be powerful, and, if they start abruptly, so much the better (and worse). A jumble of many tones is another characteristic.

In terms of telecommunications and engineering, noise has a slightly different connotation: meaninglessness. Hence, signal to noise ratio is a measure of the quality, or the information content, of a signal. This definition of noise is in some senses at odds with everyday usage: someone shouting and swearing at you may well be noisy but is unlikely to be meaningless, and the same is true of the noise of, say, gunfire. Here again, the content of the noise itself is of less importance than the reason for it.

In environmental contexts, noise is a form of pollution. Though the actual examples of it are still covered by the concept of sound out of place, this classification makes tackling noise very much easier, partly because pollution is, by definition, something to be avoided, controlled, or eliminated, and partly because this usefully lumps noise with other pollutants such as chemical wastes, excess light, radioactivity, and so on. As we shall see, one key way to deal with noise is to battle against it alongside other things. Finally, the identification of noise as a pollutant reminds one that it is not just a problem for *people*: in the ocean in particular, noise impact on marine life has come to prominence in recent years.

Though the word 'noise' is rarely used in the context of pieces of music except by people who dislike them, the intimately related subject of dissonance has been key to the development of music through history—dissonance here being defined in various ways, but always in contrast with harmony. Like noise in the more general sense, dissonance has both physical and perceptual elements. It is made by playing together (or in quick succession) notes whose frequencies are not whole-number multiples of each other, and psychologically it gives rise to a somewhat disturbing quality—but without it music would be intolerably bland.

Dissonance is not the only type of noise that is not 'out of place'. Such phenomena as shock waves and high-intensity sounds, particularly in the high-frequency region, are of enormous use in scanning everything from unborn babies to railway tracks and at higher energies are used to cut metal and to destroy tumours. Stepping down the pitch to an audible level, we arrive at the world of military acoustics, where noise weapons are of growing significance. At lower frequencies still is the strange and still incompletely understood realm of infrasound.

The derivation of the *word* 'noise' is of little help in defining it. Oddly enough it comes from 'nausea'—oddly but maybe not inappropriately: according to a 2007 survey carried out by Trevor Cox of Salford Acoustics, vomiting is the world's most unpopular noise.[2]

But, other people asked: 'Why are you writing a *HISTORY* of noise?' The answer is that the ways in which troublesome noise is viewed, and the reasons it can be so hard to control, are often rooted in the historical development of our relation to it. Noise has been humanity's permanent companion, sometimes an enemy to be battled, sometimes a servant to be trained, an element to be sprinkled sparingly in musical composition, a mystery to be solved, or a power to be propitiated, and looking at the history of noise is in some sense a way of looking at the history of ourselves.

And then, of course, some rather direct people asked (or looked as if they might like to): 'Why are *YOU* writing a book about the history of noise?' Because I have only ever had one proper job: working in the Acoustics group (for some years as its head) at the National Physical Laboratory (NPL) of the UK, in Teddington, and since then I have continued to work on a freelance basis with the subject. However, while I am a scientist, this is not a textbook; it is the first and only history of noise in all its forms, but it is by no means the only noise book; which brings me to one other reason for writing it: all the other books on the subject are technical tomes or calls to arms to fight the battle against noise. But, though there is plenty of science here—and plenty of opinion too, of course—the primary point of this book is to tell a story.

Noise is, of course, a worldwide issue and always has been, but, though the original intention of this book was to touch on noise in all periods and places, that soon proved impossible. For ancient material, there is little available outside the usual suspects: Greeks, Romans, and a few Egyptians. The dark ages are pretty much as silent as they are empty of other record. And then, until the last few decades, practically all the discussions of noise were confined very largely to the UK and to the USA—and, even there, there is an enormous preponderance of New York- and London-based material. However, to some extent at least, it seems very likely that the issues faced in these cities are, and have been, universal ones.

This also cannot be a truly scientific history: since the first recording devices were invented only in the second half of the nineteenth century and the first effective noise-measuring devices not until the 1920s, we must rely before then on many sources for evidence about noise: diaries, books, court records, and even paintings provide us with insights into the noises of their times.

Before we plunge into prehistory, we need to skim through science: noise is sound, but what is that? And what happens when it reaches our ears?

1

THE NATURE OF NOISE

A sound is not unlike the circles of ripples that spread from a stone thrown into a pool, all moving outwards from the point of impact. Each new ripple disturbs the water a little less than the one before, and, as each spreads, its speed remains the same, but its height reduces, until eventually the pool is flat and still again. Any physicists watching the pool might see things rather differently: first, the act of throwing transfers enough energy to the stone to make it fly through the air. Then, when it strikes the water surface, the stone slows, giving up some of this energy. Some goes into generating the ripples, while the rest is used to make the sound of the splash and to heat the water up a little. Though the only visible effect is the set of surface ripples, energy is now spreading through the whole pool: each circle of ripples is the edge of a growing hemispherical shell, warming the water a little more as it passes and fades.

A sudden, short sound (an *impulse*) like a clap works in the same way as the pebble. Energy spreads from the clapped hands in the form of a series of spherical sound waves, travelling about 100 times faster than water ripples. These sound waves are areas of increased pressure: as each wave reaches a new part of the air, the molecules it encounters there move closer together for a tiny fraction of a second, and then move apart again as the wave passes. A louder clap does not make faster waves than a quiet one: it just squeezes the air molecules harder, forcing them closer together and making the pressure jump higher.

Like ripples, the sound waves get weaker as they get larger and they make the room slightly hotter as they spread and die away: *very* slightly—a little energy goes a very long way when it is used to make a noise, which is why there is so much of the stuff around.

If there is an ear in the room, the first part of it that the sound waves encounter is the pinna—the external part. It used to be thought that the pinna amplifies sounds like a hearing trumpet, but it's the wrong shape for that, as well as being too soft. But it does allow us to hear much better than a hole in the head. A simple abrupt opening in a smooth, flattish surface like the scalp actually reflects most sounds, rather than allowing them in, but the folds of the pinna provide a gradual change from scalp to ear canal, allowing sound waves to enter the head with little resistance.

Once safely in, the sound waves travel down the canal (or meatus) passing the hair and wax that are there to keep flies and gnats (and earwigs, presumably) out. About 2 centimetres later, they encounter the eardrum. Just like any other sort of drum, the eardrum is a roundish disc of thin material, stretched taut so that it can easily be made to tremble—which the sound wave promptly does.

But there is nothing to hear yet—first, the tiny trembling of the eardrum is enhanced, thanks to an amazing series of structures. The eardrum, as the song so nearly goes, is connected to an earbone. And that earbone is connected to another earbone. And that earbone is connected to *another* earbone. And that earbone is connected to a . . . window, oddly enough. The bones are the smallest in the human body, and they are called the hammer (malleus), anvil (incus), and stirrup (stapes), because they look like such things (if you use your imagination in the case of the anvil) (see Fig. 1).

These earbones, or ossicles, evolved from elements of reptilian jawbones millions of years ago, and are now a series of levers that together increase the tiny trembling at the eardrum to a rather less tiny one at the window. The amplification that these bones provide

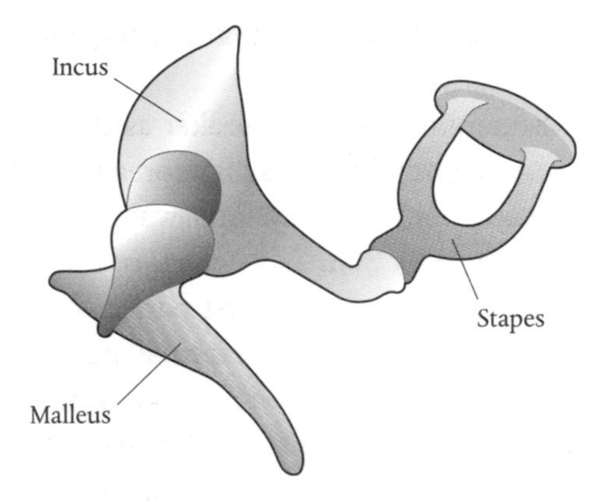

FIGURE 1. The ossicles.

allows us to hear sounds so weak that they move the eardrum less than the diameter of a single atom. (In terms of pressure, this is about one-billionth of an atmosphere—in terms of power, it would take such a source about 100 million years to boil enough water to make a small cup of tea.) If we could hear quieter sounds still, we would live in a world of continual noise, because the omnipresent random motion of air molecules would be audible. Our hearing really could not get any better.

Or at least, for those of us with undamaged hearing mechanisms it couldn't: that is, those under 16 or so who don't listen to loud MP3 players too often. For the rest of us, our hearing mechanisms have been well and truly battered by the effects of time and noise: hair cells die off as a natural part of the process of ageing, resulting in hearing loss, especially at higher frequencies. This effect can also be caused by noisy living, but it is not clear how many people's hearing loss is due to this cause. However, it is known that 120 million people in the world have hearing loss sufficient to disable them, and that all adults are deaf to sounds above about 15,000 cycles a second (15 kilohertz, abbreviated kHz).

Undamaged hearing systems can detect a wide range of frequencies: from 20 cycles a second (or hertz) to 20,000—over nine octaves.

In comparison, the range of light frequencies that we can detect, from deepest red to darkest violet, is barely one octave (that is to say, the frequency of violet light is about twice that of deep red[1]).

The range of sound energies we can cope with is even larger than the range of frequencies—in fact, it is little short of amazing. In a sense, the highest level of sound we can detect is one powerful enough to be fatal, but conventionally the upper limit is placed at the point when the experience of listening becomes one of pain (in the form of a disturbing sensation of tickling deep in the ear). A jet engine about 25 metres distant, or a pneumatic drill about a metre away, would generate such a level.

The range of energies equivalent to the difference between the quietest and loudest sounds we can hear is hard to grasp: if the thickness of this book was equivalent (in some bizarre way) to the quietest sound, the loudest would be a pile of these books high enough to reach the Moon. Or, to put it another way, if there were a sound source at ground level that emitted a tone at 500 Hz, at such a volume that it were just loud enough to be painful to listen to at a distance of a few centimetres, then one would have to travel vertically upwards to a height of over 400 kilometres before it would fade into inaudibility (assuming that the air were the same density all the way up).

To help the hearing mechanism survive exposure to powerful sounds, a muscle attached to the stirrup automatically pulls it away from the window, while another tautens the eardrum. The muscles remain tense for some time after the loud sound has stopped, which is why sounds can seem muffled after a noisy experience. This so-called acoustic reflex kicks in about 30–40 milliseconds after the sound has arrived, and takes full effect after around 200 milliseconds. Which is really rather a pity, since a sudden sound will have had time to deliver a considerable amount of damaging energy to the inner ear by then. If we had evolved on a planet where sudden loud natural sounds were more prevalent, it's likely that this response would be a bit quicker off the mark.[2]

What happens to sound waves until they reach the final ossicle is well understood, but the next stages are more doubtful. The oval window, which the stirrup connects to, is the end of an organ called the cochlea, which is about the shape and volume of a small snail. Its structure is like a fluid-filled tube that has been folded in half and then coiled up. It is in the area where the two halves of the tube are in contact that the reception of sound waves occurs. This is called the basilar membrane, and it is covered in tiny hairs.

The oval window allows sound waves to enter one end of the tube, and they then move rapidly along it, first spiralling inwards, then passing the folded-over section and spiralling outwards again. When they reach the end of the tube, a second window—round this time—bulges outwards, absorbing their energy to prevent them from returning back down the tube to cause standing waves and confusion. When sound waves impinge on the hairs of the basilar membrane, the cells at the hairs' roots register the movements by sending signals to the brain. Waves of different lengths move hairs on different parts of the membrane to different extents, so the pattern of nerve impulses that the brain receives allows the form of the original sound wave to be decoded. The more distant parts of the membrane from the eardrum respond to lower frequency sounds. An important attribute of the membrane is that, if two sounds of similar frequencies impinge on the same area, the louder of the two will tend to mask the quieter one. If two sounds impinge on different areas, they can be heard separately, even if they are very different in loudness—which is why we can hear so well in noisy conditions if the noise is not at speech frequencies (and why, if it is, we pitch our voices to move them to a different frequency range).

This is only one of our hearing systems. Even without ear canals or eardrums, we would still be able to hear, thanks to bone conduction. In this system, sound vibrations travel through the mastoid bone behind the ear, and are conducted directly to the cochlea.

A general view of the ear's structure is shown in Fig. 2a, while Fig. 2b is a schematic diagram.

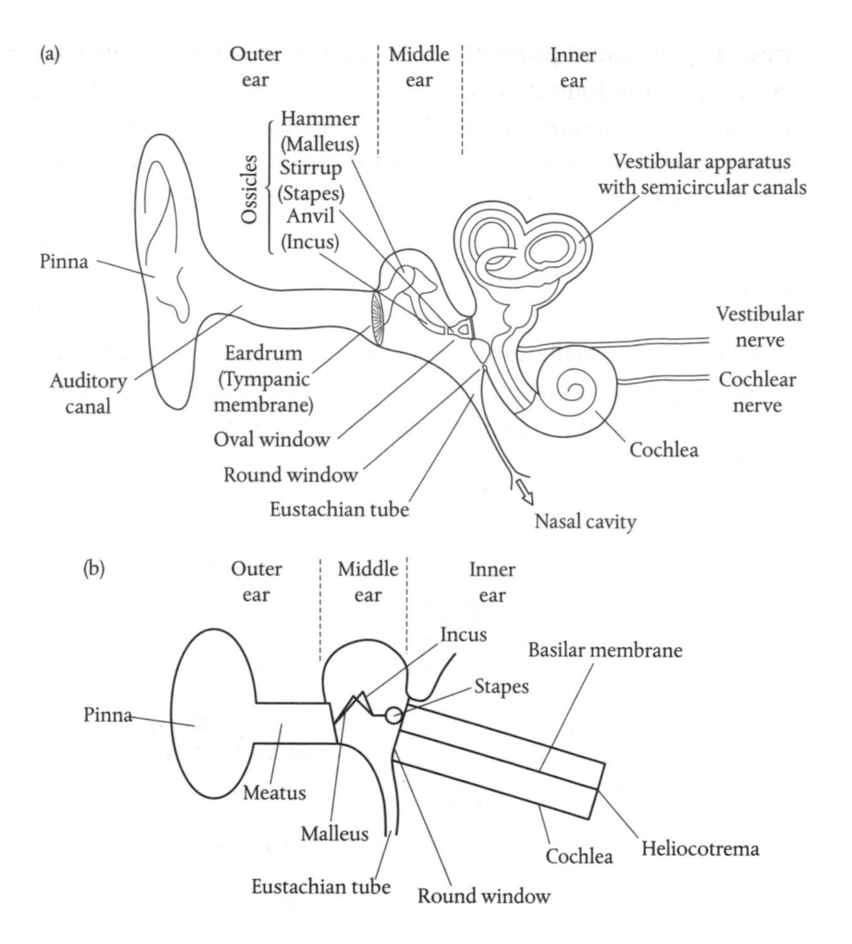

FIGURE 2a and 2b. The mechanism of the ear. The sizes of the middle and inner ear are greatly exaggerated.

Where noise is concerned, the amount of sound is clearly a key factor, but what should be measured, how to do so, and what units to express the answer in are not so obvious. The development of measurements and units of sound is part of the story this book will tell, but a brief summary is as follows.

Noise, like any other type of sound, can be measured in terms of sound *pressure* (the thing that most sound-measuring devices measure) or sound *power* (the total energy sent out—often more relevant to the impact of sound but trickier to measure). Sound *intensity* is the amount of sound power in a particular area. Sadly, none of these quantities is equivalent to the loudness (sometimes called 'perceived' loudness, but that is really a tautology) of a sound—and, even more sadly, no instrument can measure this properly. Fortunately, there are approximate relations between these quantities. (The term 'volume' is roughly equivalent to loudness but is not usually used except on dials on audio equipment).

Sound pressure, power, or intensity can be measured in various units but are most frequently referred to in decibels, often used to label charts like Fig. 3.

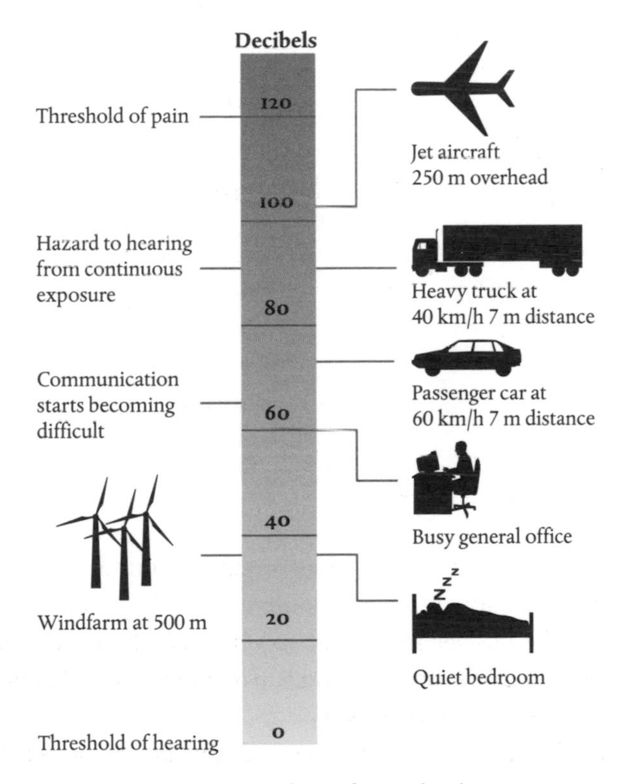

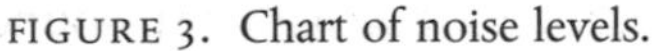
FIGURE 3. Chart of noise levels.

Decibels will be defined later, but the key point to bear in mind about them is that they are not linear measurements but logarithmic: as a sound gets louder, what happens is as follows.

- A 3 dB increase means a doubling of sound intensity and of sound power.
- A 6 dB increase means a doubling of sound pressure.
- A 10 dB increase means an approximate doubling of loudness or 'volume'.

Conversely, if your sound-measuring device tells you that the sound pressure has increased by 20 dB, that means that

- the sound pressure (and the voltage your device measures) has got 10 times larger and
- the sound power (and intensity) has got about 100 times greater while
- the loudness (and volume) has got about 4 times greater.

(There are all sorts of assumptions here—in particular, that the sound does not change in frequency as it increases and that it radiates equally in all directions.)

To look at this from a third and final perspective: it is pretty much impossible to hear that a sound has changed in pressure by 1 dB. A 3 dB change can just about be detected by the ear (which is actually none too impressive, considering that this represents a doubling of the sound energy impinging on your eardrum). A 5 dB change is clearly noticeable, and, as mentioned above, a 10 dB change will be about twice as loud and a 20 dB change about four times as loud.

2

A SILENT BANG

Despite a promising name, the Big Bang was silent—a sudden burst of energy in which time and space began, forming the Universe as it spread. With no space to expand into, there could be no medium around it into which sound waves could possibly propagate. But, in cosmic terms, the Universe was not silent for long—380,000 years later (a mere 0.003 per cent of its present age), it was filled with sound. And, this was not the random roar of white noise that one might perhaps expect—it was a sound with a pitch: it had a characteristic wavelength.

It would not, however, have been an audible sound to any eared creatures, could they have existed so far back in time, before even the stars were born: a vast object like the Universe makes a very low sound indeed—about one-trillionth of a hertz.

The reason that there was such a vast deep tone in the infancy of space and time is closely connected to one of the most mysterious and important aspects of the Universe's history: structure, of which sound is a signpost. If the Universe had remained as it began, a completely homogenous, smoothed-out volume of energy, then galaxies, stars, and people could not exist today. But, for reasons that are still unclear, there was a clumpiness in the early Universe—some areas were a little denser than others, and it was these denser areas that would eventually become stars and galaxies. Density means gravity, and this gravity attracted nearby matter (then in the form of plasma—a 'gas' of ions). The motion of that matter caused compression, heating the plasma, which in turn increased its output of radiation. The force of this radiation counteracted the gravitational force, and so the compression became an expansion—and it is this cycle of compression and expansion that formed the primordial sound waves.[1]

The wavelengths of the wail of the baby Universe—measured in hundreds of thousands of light years—were limited by the speed with which the pull of gravity travelled from one region to another, which is the speed of light. So, there was an ever-falling lowest possible pitch to the Universe, and consequently a gradually descending tone marked its growth.

The variation in pressure of the sound was around 1 per cent, or 110 dB, the kind of level that would be associated with motorway traffic a few metres away and over thirteen billion years later.

In the early Universe, as new generations of stars formed using the nuclear reaction products of the old, planets like ours formed with them—and sound waves surged and echoed through their structures and their atmospheres and, later, their hydrospheres too. But, as far as we know, for ten billion years there was nothing to hear them.

Then, around four billion years ago, life started in the seas of Earth, and it was there that the first hearing mechanisms evolved. The fossil record has left no trace in the rocks of quite when, or in which creatures, these first developments arose—perhaps weedlike organisms clinging to rocks were the first to sense the tremble of stone as waves crashed overhead, or perhaps it was some ancestral jellyfish that vibrated in sympathy with the sound waves that passed through it and the water in which it drifted.

The ability to detect sound and to react to it—at first, perhaps simply as a warning of danger—must have been a significant evolutionary advantage, and so the creatures with the best detecting systems survived and multiplied. Over many generations, first specialized cells, and then whole organs, developed, with the ability not only to detect the presence of sound but to distinguish different pitches too. With two separate sensors, the direction of the sound could be determined. Gradually, as the blind forces of nature experimented with different structures, special amplifying mechanisms appeared, increasing the sensitivity of the systems. The flow of sensory information concerning sound grew in complexity, and the receiving brains—or what passed for them—changed too.

By the time the first amphibians crawled from the warm seas about 400 million years ago, they were equipped with complicated structures that could detect sound waves both underwater and in air, and determine their strengths, pitches, and directions. These sound waves allowed prey to be tracked, young to be guided, mates to be attracted, and danger to be anticipated. These creatures could feel sound vibrations too.

An isolated glimpse of the state of one hearing mechanism in those far distant times was provided in 2007 by the discovery of the well-preserved jawbone of a primitive mammal called *Yanoconodon allini*. While it has stirrup, anvil, and hammer bones like ours, a fourth bone is also present, which connects the other bones to Yanoconodon's jaw (see Fig. 4). This link allowed Yanoconodon to hear both sound waves and ground-borne vibrations. In modern humans, the only trace of this bone is Meckel's cartilage, a tiny strip that disappears before birth.

Since fossil evidence is very limited, insights into the evolution of hearing have been provided through studies of contemporary species like opossums, which are believed to be similar to prehistoric creatures. Hearing tests on a range of such animals suggest that high-frequency hearing is a peculiarly mammalian ability, and that it is found most often in those mammals with close-set ears. The reason for this is thought to be that having such ears allows the location of higher-frequency sound sources to be located accurately—very useful when the sound source in question is some tasty, or deadly, animal.

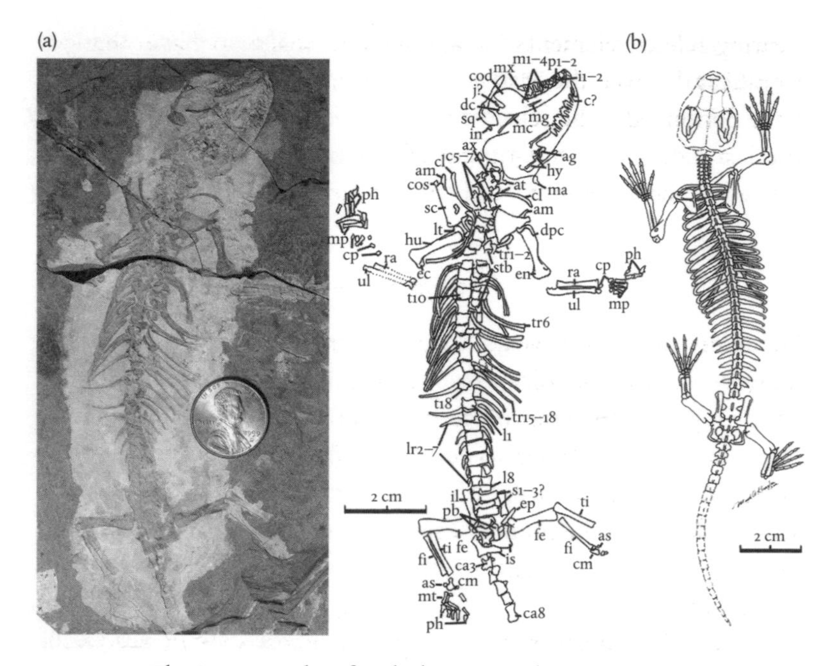

FIGURE 4. The Yanoconodon fossil, diagram and reconstruction; *ma* is the malleus (hammer) and *mc* is Meckel's cartilage.

From Z.-X. Luo, P. Chen, G. Li, and M. Chen, 'A New Eutriconodont Mammal and Evolutionary Development in Early Mammals', *Nature*, 446 (2007), 288–93; reprinted by permission of Macmillan Publishers Ltd.

Until sometime in the Eocene epoch (between about thirty-four million and fifty-six million years ago), the hearing range of our own ancestors seems gradually to have increased, as the hearing mechanism improved its low-frequency performance. Since then, a gradual reduction in sensitivity to higher frequencies has meant that the range of audible frequencies has contracted.

Comparisons of human genes with those of chimpanzees and other close relatives have shown that a gene called alpha tectorin has had an especially high rate of mutation, suggesting that it confers an important evolutionary advantage. Alpha tectorin determines some of the acoustic properties of the inner ear. Meanwhile, genetic comparisons of different human populations show that the hearing-related elements of our genetic make-up have changed significantly over the last 50,000 years, and that hearing genes have continued to be selected by evolutionary pressures even over the last 2,000. This fine-tuning of our hearing ability is probably related to the development of spoken language.

Ancient noise

For many millions of years, loud, irregular sounds have signalled danger—the roar of volcanoes, the crackle of lightning, the rumble of earthquakes, the cries of hungry enemies. And instinctive reactions of fear and anger to loud sounds remain with us today, defining for us the concept of noise. Our tiny shrewlike ancestors, scampering from the thunder of tyrannosaurs, are connected by an unbroken thread to the angry restless sleepers under the flight paths of aircraft.

The first humans were hunters and gatherers, moving across the great temperate plains of the world in search of prey. Their acoustic environment is almost unimaginably different from ours: there must have been far fewer repetitive sounds, and, except near the ocean or fast rivers, continuous sound was rare too. (It is said that nineteenth-century Native Americans never made camp near running water, as it might have masked the sound of enemies. Perhaps early hunter-gatherers were the same.) What sounds there were must have been of great interest to our long-distant forebears—in general, high-pitched cries must have suggested small prey, while the louder, lower-pitched sounds made by larger animals must have been heard with more trepidation. Sounds of most sorts in those times really mattered—some were literally a matter of life and death, and all must have commanded attention. It is the attention-getting nature of sound that persists to this day and that accounts for the impossibility of simply switching off the outside acoustic world—our lack of 'ear-lids'—even when we sleep. In fact, so impossible is it to ignore sounds entirely that the brain's electrical responses to sounds are used to indicate the depth of unconsciousness during surgery.

Even quiet sounds can become frightening noises if they are unidentifiable, particularly when heard in the quiet of the night when their source remains unseen. Such reactions must go back to eras long before human history began, and be deeply hard-wired into our brains through the steady process of evolution: ancestors that cowered in fear from sounds that came from beyond the safe circle of firelight must often have survived, while braver souls who ventured beyond never returned to breed and pass on their foolhardy courage. It is not surprising then that unknown noises, when heard in darkness—things that go bump in the night, or startle audiences in shadowy cinemas—are still such a potent source of fear.

As far as artificial noise sources are concerned, perhaps it was not until early humans began to chip flint to make tools about 2.5 million years ago that anything like a modern industrial noise was made. Maybe it was even the first human-made noise, in the sense of something that annoys.

Noise in the sense of dissonance, together with its complement, harmony, also has its origins long before recorded history. Some archaeologists believe that music pre-dates humanity entirely, with evidence that Neanderthals made musical instruments, but the earliest definite evidence is in the form of bone and ivory flutes, found in south-western Germany. They date back more than 35,000 years, to the middle Paleolithic period, long before even the most primitive towns existed.

Despite the long love affair between humans and music, it is still highly mysterious: we are hardly further forward today than Charles Darwin was in 1871, when he confessed himself baffled by its evolutionary function. What is clear is that there are qualities to musical pieces that transcend cultural differences and backgrounds. In 2009, an experiment by the Max Planck Institute in Leipzig found that Mafa tribe members, from Cameroon, had the same sorts of emotional response to classical piano pieces that a Western audience did.[2] One other recent discovery is that music affects the premotor parts of the brain, the function of which is to prime us for physical activity.[3] In other words, music makes us want to move, whether by working together or by performing together.

By 20,000 years ago, our cave-dwelling ancestors were already filling their environments with controlled sound: rock-gongs (called lithophones) have been found in caves in many parts of the world: we know the approximate dates of use of some of the gongs—in the Pyrenees, for example—because layers of calcium salts have had time to build up over the ancient traces of well-aimed impacts.

It may well be that in the prehistoric past sound and noise had a more prominent and significant role than they do today. The greater darkness both of the night and of internal spaces would have left our ancestors relying more on their ears than their eyes for much of the time—or all the time, for all those who shared our many vision defects, thousands of years before artificial glass, let alone glasses, had been invented. And, of course, before the invention of writing, the only records were spoken ones. Finally, so long as the world remained a quiet one, sounds that did occur would have attracted more attention, simply through their rarity.

In around 10,000 BCE the first farming settlements appeared in the areas between the Tigris and Euphrates rivers known as Mesopotamia, and from those settlements the first cities grew. From cities arose the sounds not only of crowds but also of trades: in addition to its defensive advantages, one of the key benefits of living in a settlement is that not everyone has to do everything—so people can be potters or farmers, builders or artists, bartering the products of their skills for those of others. Since these trades each had a distinctive sound and since people with similar jobs often worked in the same areas, different zones of noise would quickly have developed. It has been suggested that, by the ninth century, the Huns were constructing their communities in concentric circles—as many as nine in some cases—for acoustic reasons. Each circle was a rampart, and, between them, farms and small groups of dwellings were arranged so that each was within easy shouting distance of the next—while the distances between the circles was such that trumpets could be used to signal between them.[4] There are indeed sites built with this kind of structure, which would certainly have allowed such hearable links to be made—but whether the acoustics were really the impetus for the design is another matter. This is the key problem that dogs evidence purporting to reveal acoustically based design of ancient structures: they can often generate impressive sounds, but were they deliberately engineered to do so? For instance, a 1990s study of six ancient underground constructions (burial chambers and temples) showed that all of them had resonances around 100 Hz—which would magnify nicely the voices of men who stood at particular locations within them. [5] One can well imagine the impact of such effects in such dark and holy places, often populated by many more rotting and skeletonized bodies than living ones. But whether the builders knew what they were doing, or simply exploited the resonances they found, is impossible to establish.

One of the best pieces of evidence for acoustic design in the prehistoric world is Stonehenge.[6] While the outer sides of its stones are rough and relatively unfinished, the inner faces are much smoother and slightly concave. The effect of this is to make the space inside Stonehenge nearly as reverberant as a concert hall, with a reverberation time of just over a second. Bearing in mind that the structure has no roof, this seems to be an impressive feat of acoustic engineering, though one whose effect is rather lost today, because of the removal or fall of many of the stones. When it was complete, Stonehenge must have been an ideal acoustic space, working so that a speaker's voice filled all parts of it to a similar volume. This effect is more noticeable in some of the numerous reconstructions of the structure that exist, such as that in Washington State, USA.

Whatever we think of the status of acoustic design before history began, we can at least have some appreciation of what our ancestors thought of as noise: the story of Gilgamesh, the half-divine ruler of the City of Uruk, in Mesopotamia, was written in India around 2000 BCE and set about 700 years earlier. This great epic relates how the gods punished humanity for making too much noise:

> You know the city Shurrupak, it stands on the banks of Euphrates? That city grew old and the gods that were in it were old...In those days the world teemed, the people multiplied, the world bellowed like a wild bull, and the great god was aroused by the clamour and he said to the gods in council, 'The uproar of mankind is intolerable and sleep is no longer possible by reason of the babble.' So the gods in their hearts were moved to let loose the deluge...[7]

The fact that this myth originated in one of the world's first cities shows that our modern concept of noise as unwanted sound is contemporary to—or earlier than—the birth of the city. Like an annoying teenager, Gilgamesh himself used noise to show his power—setting off alarm bells just for fun: 'But the men of Uruk muttered in their houses, "Gilgamesh sounds the tocsin for his amusement, his arrogance has no bounds by day or night."'[8]

Some have suggested that it was in Egypt, as long ago as the seventeenth century BCE, that the first medical references to noises appear, as descriptions of tinnitus and of auscultation, (listening to the body's internal sounds to ascertain its health or otherwise). There is indeed a papyrus from Crocodilopolis, written in Demotic (a late Egyptian script), which says: 'In the group of the complaints of the ear can be mentioned, raging,[9] stinging, decay and abscess, illness of the earwax gland, loss of hearing, and worms.'[10] However, this papyrus probably reflects Greco-Roman medicine at least as much as Egyptian. References to auscultation are similarly obscure.

It is not until the fifth century BCE that the first clear description of tinnitus is to be found, in the works of the Greek physician Hippocrates, who referred to it as a slight buzzing sound in the ear. Hippocrates is also the first physician known to have recommended that sick people should be accommodated in areas distant from noise sources.

It would be nice to talk here about how noise brought the walls of Jericho down, but the myth of the destruction of the walls of Jericho by the sound of Joshua's trumpeters is just that: radiocarbon measurements show that the city was destroyed around 1560 BCE (plus or minus thirty-eight years), and not in 1407 BCE, when Joshua was around. In fact, the city was deserted at that date.

The rational use of sound was still very much the exception; in general, in ancient Egypt and other early civilizations, loud noises were still mysterious and portentous. In 27 BCE an earthquake tremor struck the twin Colossi of Memnon, two massive stone statues of Pharaoh Amenhotep. Their function was to stand guard at the entrance to Amenhotep's memorial temple, which had brooded over the landscape for the past thirteen centuries. As a result of the tremor, the upper half of one of the statues collapsed and the lower half was cracked.

From then on, from time to time—but always near dawn—strange noises were heard from the battered Colossus. Some said they were like the strings of a musical instrument snapping, while others heard unearthly whistling sounds, or a brassy clashing. A few even heard words among the sounds and decided that an oracle had taken up residence. Even those who heard only wordless noises did not hesitate to ascribe them to supernatural causes (no doubt encouraged by the local priests and everyone else who might benefit from an upturn in the tourist trade). Thanks to the accounts of famous contemporaries, especially the Greek historian Strabo, who visited the site in 20 BCE, the fame of the noisy statue spread through the civilized world. Even Roman emperors came in the hope of hearing the mysterious sounds, which by then were supposed to confer good luck on the hearer. Such pilgrimages continued for over two centuries, until Emperor Septimus Severus, disappointed in being unrewarded by a sound, added some layers of sandstone to the statue in an attempt to propitiate the oracle that lurked within. The sounds were never heard again.

Mysterious sound

While harmony and dissonance were shining examples of science, noises in general were still frequently invested with a sense of mystery—and indeed were genuinely mysterious in some cases.

Singing (or booming) sands, for instance, are first mentioned in the ninth century CE, in China:

> The Hill of Sounding Sand . . . has strange supernatural qualities . . . In the height of summer the sand gives out noises of itself, and if trodden by men or horses the noise is heard by men tens of *li*[11] away . . . The ancients called this hill the Sounding Sand; they deified the sand and worshipped it there.[12]

These days, of course, we can all hear singing sands whenever we like, courtesy of YouTube. But, even divorced from its setting, the sound still conveys an eerie, mysterious quality, and the effect on ancient—and superstitious—travellers must have been disturbing indeed. And what recordings cannot give us is the eerie physical sensation that accompanies the sound: if one stands on the dune, a restless trembling can be felt, and the whole surface shifts and ripples underfoot, adding to the unsettling effects of the sound—which can itself be loud enough to drown normal speech. The noise lasts for several minutes and usually consists of a fairly narrow-frequency tone between around 50 Hz and 300 Hz, though the pitch varies from time to time and dune to dune, being determined both by particle size and by the depth of the loose upper sand layers. The effect is strongest with dry sand and rounded grains, and it can be initiated either by the desert wind or by the footsteps of a traveller. Though a number of explanations have been suggested, there is still some disagreement over the details of the mechanism. However, it seems likely that the effect is due to the avalanche-like motion of loose surface layers of sand, which generate the distinctive tones in moving across one another. Hard-packed sand below reflects the sound back upwards, enhancing the effect.

An equally strange sound, also first reported in China at about the same time as the earliest reference to singing sands, is one that occasionally accompanies meteors. The report was to a meteor 'which made a noise like a flock of cranes in flight'.[13] Occasional reports of such space-based noises continued, and a whole collection was gathered together by the astronomer Edmond Halley in the 1710s. More recently, in 2001, many observers of the November Leonid meteor shower reported the same type of noise. The baffling thing about the sounds is that they are heard simultaneously with the meteors being seen. Since meteors usually form 50 kilometres above the Earth at the very least, one would expect any sounds that they make to lag behind them by several minutes—and, to reach the Earth's surface from such a rarefied and distant part of the atmosphere at all, they would also need to be incredibly loud at their source.

The only plausible explanation for the simultaneity is that the energy travels from the meteors in the form of electromagnetic radiation and is somehow converted to sound near the observers—or possibly even inside them; it has been suggested that the radiation is in the form of microwaves, which set up resonances in the skull or inner ear and either generate actual sound waves in the head or stimulate the cochlea in such a way that the impression of sound is made. (Whether this is the true explanation of these so-called electrophonic meteors or not, sound weapons have recently been produced that exploit this very effect.) Other theories, equally speculative, suggest that the radiation from the meteors is at the other end of the radio spectrum: very long waves that can cause resonances in structures like pine needles—or even spectacle frames.

Another noise, as fearsome as it was mysterious for as far back as living things could feel either emotion, is the ominous rumble of an

FIGURE 5. Zhang Heng's earthquake detector.
Courtesy of Marilyn Shea.

earthquake. Even before the first reports of either electrophonic meteors or singing sands—but once again in China—the scholar Zhang Heng constructed a machine to detect the seismic waves that are generated along with the thunderous sounds of an earthquake. The world's first seismograph was a vessel with a ring of dragons' heads around it, each of which held a metal ball (see Fig. 5). Below each dragon a metal frog waited open-mouthed, until, when an earthquake struck, one or more balls would drop with a clang into them, indicating roughly the earthquake's direction.

Whether the detector really worked or not is unclear, since only the outer casing survives, but there is a legend that the court of the Emperor An in Luoyang, the Chinese capital at the time, was amused one day when the detector installed there registered that an earthquake was occurring. The mockery of Zhang lasted until the reports of the earthquake that caused it arrived at court—from 400 kilometres away, in the direction his detector had indicated.

The mystery and the power of sound are reflected in the starring role it frequently plays in myths of the world's creation. Of the many creation tales told in ancient Egypt, one of those originating in Memphis is the most acoustical: Ptah, the god of craftsmen, produced all the other gods and the rest of the contents of the Universe simply by speaking their names. But perhaps a more evocative myth involves a sound, rather than articulated speech: according to the myths of Heliopolis, it was the cry of the Bennu bird that began the Universe.

3

CLASSICAL NOISE

The warlike associations of noise are rooted in the distant past, and the concept of the war cry is common to many times and places, being used to rally the blood lust of the attackers to unite them into a single force, while at the same time intimidating their enemies. According to the Greek historian Polybius, the Romans well understood the impact of noise in war, as his account of a battle between Romans and Carthaginians in 255 BCE—an episode in the First Punic War—makes clear:

> No sooner had Xanthippes ordered the elephant drivers to advance and break the enemy's line and the cavalry on each wing to execute a turning movement and charge, than the Roman army, clashing their shields and spears together as is their custom, and uttering their battle-cry, advanced against the foe.[1]

But some enemies were able to beat the Romans at their own noisy game; according to Polybius again:

> The Romans, however, were on the one hand encouraged by having caught the enemy between their two armies, but on the other they were terrified by the fine order of the Celtic host and the dreadful din, for there were innumerable horn-blowers and trumpeters, and, as the whole army were shouting their war-cries at the same time, there was such a tumult of sound that it seemed that not only the trumpets and the soldiers but all the country around had got a voice and caught up the cry.'[2]

The Germans in the first century CE also made use of battle cries. According to a contemporary account:

> They also have the well-known kind of chant that they call *baritus*. By the rendering of this they not only kindle their courage, but, merely by listening to the sound, they can forecast the issue of an approaching engagement. For they either terrify their foes or themselves become frightened, according to the character of the noise they make upon the battlefield; and they regard it not merely as so many voices chanting together but as a unison of valour. What they particularly aim at is a harsh, intermittent roar; and they hold their shields in front of their mouths, so that the sound is amplified into a deeper crescendo by the reverberation.[3]

Battle cries have been a feature of war ever since. What is actually said varies widely—in nineteenth-century America, the Sioux shouted 'Hoka Hey!' ('Today is a good day to die!')—and some Sioux-descended US soldiers continue to do so to this day. In the Second World War, Finnish soldiers used the war cry 'Tulta munille!' ('Fire at their balls!'), while Japanese kamikaze pilots would shout 'Banzai!' ('Ten thousand years'). In modern times, war cries are still prevalent, but less literal, perhaps acknowledging that the point of the cry is its noise, not its content, with American soldiers in training nowadays all using some variation on 'Hoo-YAH!'

A notable and noisy source of warlike sound in the ancient world was the trumpet, and early references to it make clear it was used as a warning device in battle. In Homer's *Iliad*, written in about 850 BCE, the role of the trumpet in war is mentioned, along with the idea that the power of a great warrior can be expressed through his voice, the power of which alone (ideally with a little help from a friendly goddess) was enough to subdue those who heard it. Homer describes the scene:

> There did [Achilles] stand and shout aloud. Minerva also raised her voice from afar, and spread terror unspeakable among the Trojans. Ringing as the note of a trumpet that sounds alarm when the foe is at

> the gates of a city, even so brazen was the voice of the son of Aeacus, and when the Trojans heard its clarion tones they were dismayed; the horses turned back with their chariots for they boded mischief, and their drivers were awe-struck by the steady flame which the grey-eyed goddess had kindled above the head of the great son of Peleus.
>
> Thrice did Achilles raise his loud cry as he stood by the trench, and thrice were the Trojans and their brave allies thrown into confusion; whereon twelve of their noblest champions fell beneath the wheels of their chariots and perished by their own spears.[4]

Trumpets became so popular in the ancient world that a trumpeters' contest was introduced into the Olympic games in 396 BCE; the winners of the competition were then allowed to use their trumpets to announce the beginning of all the other events. Among the champion trumpeters was Herodoros of Megara, who won ten championships in a row (a phenomenal achievement, given the thirty-six-year time span involved in an age when few people even survived until their fifties). Herodoros was supposed to be such a skilled and powerful trumpeter that he could blow two trumpets at once, and, legend has it, the rousing effect of this feat was sufficient for the army of King Demetrios Poliorketes to defeat the citizens of Argos in battle.

In most places, for many centuries, trumpets and horns were the main acoustical instruments of war, but, as far as European ears were concerned, something new was added to the mix in 1055 CE, at a battle between Moors and Christians near Badajoz, Spain:

> there was a sound which had never before been heard in the armies of Europe: drums. The drummers of Africa built up a thunderous roar, and the sound was not without effect on the Christians; some of them thought the earth was shaking . . . They were disastrously defeated, and King Alfonso himself barely escaped. That same night Yusuf had all the Christian corpses beheaded and the heads piled up into hills, from whose tops at dawn his muezzins summoned the troops to prayer.[5]

The secrets of sound

The first serious considerations of sound as a natural phenomenon that was accessible to human investigation were those of the ancient Greeks, a group of people who, perhaps for the first time in history, felt able to speculate on the nature of the world relatively unshackled by the chains of dogma or the fear of persecution from religious authority.

The first person known to have thought about sound in this way was Pythagoras, and his analysis was to have far-reaching consequences, not just for the study of sound and noise, but for the whole of physical science thereafter. Pythagoras' route to the foundations of science began with discordant sound and the concept of dissonance. By his time, musical instruments of many kinds were in existence, and the bases of their operation were understood: the lyre, for example, was said to have been invented by Hermes because he realized that the shell of a turtle could be used as a resonator. And it was a matter of common experience that some notes sounded pleasant when heard together (or rather, according to surviving Greek writings, in succession), while other combinations did not. In particular, the consonances[6] we know today as the octave, the fifth, and the fourth were well known to musicians of his time. What Pythagoras did was to relate such consonances to simple ratios of whole numbers. His great success in doing this not only began the scientific study of music but also encouraged him and later thinkers to view number as the key to the way the Universe works. And, just as Pythagoras' search for truth was guided by the criteria of mathematical simplicity and literally of harmony, so scientists ever since have found that truth is most reliably tracked down through theories that are as simple, elegant, and harmonious as possible; as Einstein put it, scientific research is based on 'admiration for the beauty and belief in the logical simplicity of the order and harmony that we can grasp humbly and only imperfectly'.[7] The word 'rational' itself comes from Pythagoras' powerful idea that whole number ratios hold the secrets of science.

It would be fascinating to know more about how Pythagoras, one of the founding fathers of science in general and acoustics in particular, developed his theories, but in fact practically nothing is known about him, so obsessed was he with secrecy. His followers formed a sect that jealously guarded the bases of its beliefs, so practically all we know about Pythagorean science is a probably garbled version of some of its conclusions, not its methods. Its precepts included such rules as 'do not sit on a quart pot', 'eat no beans', and 'do not allow swallows to nest in your house'.

One of the few anecdotes relating to Pythagoras is that he was as impressed with the sensual qualities of sound as the scientific ones. According to Censorius Datianus, a Roman Politician who lived centuries later, he 'kept a lyre with him to make music before going to sleep and upon waking, in order always to fill his soul with its divine quality'.[8]

So science, and especially acoustics, were off to a promising start—but what Pythagoras did next is typical of the weakness that dogs the Greek approach to scientific enquiry: taking a simple theory, valid—if at all—in a limited area of application, and extrapolating a vast and elaborate universal framework from it. In Pythagoras' case, the scheme he developed was one that was to linger on for thousands of years. As usual, we have to rely on later writers for information about it, in this case Aristotle, who says:

> Some thinkers suppose that the motions of bodies of that size [i.e., planets] must produce a noise, since on our Earth the motion of bodies far inferior in size and in speed of movement has that effect. Also, when the Sun and the Moon, they say, and all the stars, so great in number and in size, are moving with so rapid a motion, how should they not produce a sound immensely great? Starting from this argument, and from the observation that their speeds, as measured by their distances, are in the same ratio as musical concordances,

> they assert that the sound given forth by the circular movement of the stars is a harmony...
>
> It appears unaccountable that we should not hear this music, but they explain this by saying that the sound is in our ears from the moment of birth and is thus indistinguishable from its contrary silence, since sound and silence are discriminated by mutual contrast.[9]

Alexander of Aphrodisias, who lived in the early third century CE, added that, 'the sounds which they make during this motion being deep in the case of the slower, and high in the case of the quicker, these sounds then, depending on the ratio of the distances, are such that their combined effect is harmonious'.[10]

Though the idea was itself mistaken, the Pythagoreans' concept of unheard sounds was remarkably prescient. As Gottfried Leibniz pointed out over a thousand years later: 'it is well to make distinction between the perception, which is the inner state... representing external things, and apperception, which is consciousness or the reflective knowledge of this inner state...'.[11]

In other words, we sometimes become aware of a sound or other sensation only when we notice (using our 'apperception') that we have stopped hearing it (an example often given is that of a ticking clock, which is noticed only when it falls silent). This type of experience shows that, even though consciously we may be unaware of sounds, they continue to be 'noticed' in some sense by the unconscious. This observation is very relevant to background noise, which can have effects on people even though they are not consciously aware of its presence.

Despite the lack of evidence to support it, the idea of the unhearable musical harmony of the spheres caught on, and remained popular for centuries. In *The Merchant of Venice,* for instance, Shakespeare says:

> Look, how the floor of heaven
> Is thick inlaid with patines of bright gold;

> There's not the smallest orb which thou behold'st
> But in his motion like an angel sings...
> Such harmony is in immortal souls;
> But, whilst this muddy vesture of decay
> Doth grossly close it in, we cannot hear it. (v. I, 56–63)

It was not simply the enthusiasm for wild extrapolation that handicapped the development of Greek science, but a distaste for experimental confirmation of theories. That is not, however, what one might think from reading the accounts of early historians, who make it sound as if Pythagoras' acoustical discoveries were made in a highly empirical way: supposedly, he found out by experiment that, when an anvil is struck by pairs of hammers, the sounds produced were an octave apart when the hammers' weights were in the ratio 2:1. Similarly, the consonance that we call the fourth could be made when one hammer weighed four-thirds as much as the other, and so on.

This remained the official account for centuries, despite the fact that it cannot possibly be correct: simple ratios of hammer weight just don't give consonances of any sort. There are plenty of simple acoustical demonstrations that do the job nicely (pairs of strings whose lengths are described by such ratios, for instance), but not this one.

What is surprising is not so much that Pythagoras' experiments—if there were any—were not correctly reported, but that, in all the centuries after him, no one actually thought to check whether they worked. This type of limitation was to plague all areas of science right into the seventeenth century, when heated debates about why the weight of a bowl of water does not increase when a goldfish is put in it continued undisturbed by a single experiment—which would have shown that it *does*.

In ancient Greece, the absence of an empirical approach as part of the route to knowledge was the result of a conscious rejection of it: Plato, for example, was clear that only reason, pure and unmodified by experiment, was acceptable. In terms of acoustical experimentation in particular, he was uncompromising, censuring 'those gentlemen who tease mad tortured strings and rack them on the pegs of the instrument... setting their ears before their understanding'.[12]

So, Plato insisted, reason was to be followed wherever it led. In acoustics, this is where it got him:

> In considering the third kind of sense, hearing, we must speak of the causes in which it originates. We may in general assume sound to be a blow which passes through the ears, and is transmitted by means of the air, the brain, and the blood, to the soul, and that hearing is the vibration of this blow, which begins in the head and ends in the region of the liver. The sound which moves swiftly is acute, and the sound which moves slowly is grave, and that which is regular is equable and smooth, and the reverse is harsh. A great body of sound is loud, and a small body of sound the reverse.[13]

Clearly there are nuggets of truth mixed in here with such startling ideas as that the liver is the physical end point of heard sounds, and Aristotle came to similarly mixed conclusions, deciding that 'sound is heard [first] in air and then more faintly in water'. Not a great start, considering that waterborne sounds travel much faster than airborne ones; but he goes on to say:

> It is not the air or the water, however, which chiefly determine the production of sound; on the contrary, there must be solid bodies colliding with one another and with the air; and this happens when the air after being struck resists the impact and is not dispersed. Hence the air must be struck quickly and forcibly if it is to give forth sound; for the movement of the striker must be too rapid to allow the air time to disperse.[14]

A large dose of caution is essential in reading much into what the ancient Greeks wrote, even apart from the minefields of translation and the problem that many accounts are second-hand at the very best. The lack of experimental tests meant that no hypotheses were either definitely rejected, or well corroborated, with the result that a great many conflicting theories were in uneasy simultaneous existence. In the area of sound, though the Greeks were clear that air or water must 'carry' it in some way, there was no consensus as to whether the air or the water itself must move from the source to the ear. The nature of frequency was also only vaguely appreciated—high frequencies meant high speeds, but speeds of *what* was not decided. Even the relation of frequency to pitch, so commonplace to us, was only really enunciated centuries later by Galileo, and this persistent vagueness led to some very odd statements, like Anaxagoras': 'Large animals hear loud sounds and sounds far away, and the more minute sounds escape them; while small animals hear sounds that are minute and close at hand.'[15]

Noise in action

Despite the lack of progress in theoretical acoustics, the practicalities of the subject were already proving useful, with such handy tips as: 'In sawing through a bar pour on oil; for thus the sawing will go faster and with less noise. And if a sponge be tied to the saw and to the bar, the noise will be much less distinct'—that is from Aeneas Tacitus, writing sometime around the fourth century BCE.[16] Tacitus was a writer on military matters, and it is in this area that we know most about the Greeks' use of noise. For instance, in the 500s BCE, the Persians, as part of their long military campaign against the Greeks, laid siege to the Libyan town of Barca. After a tedious siege, the Persians tried tunnelling their way into the city. Herodotus, the celebrated (if not very reliable) Greek historian, tells us what happened next:

> their mines were discovered by a man who was a worker in brass, who went with a brazen shield all round the fortress, and laid it on the ground inside the city. In other places the shield, when he laid it down, was quite dumb; but where the ground was undermined, there

> the brass of the shield rang. Here, therefore, the Barcaeans countermined, and slew the Persian diggers. Such was the way in which the mines were discovered.[17]

Presumably the shield was 'laid down' quite forcefully.

This use of sound seems to have spread. In 214 BCE Philip V of Macedon's troops were engaged in another siege, and this time the Illyrian town of Appolonia was the unlucky target. Spies informed the Apollonian generals that the Macedonians were digging tunnels in preparation for a secret attack on the city. Acting swiftly on this intelligence,

> Trypho of Alexandria, who was the architect to the city, made several excavations within the wall, and, digging through, advanced an arrow's flight beyond the walls. In these excavations he suspended brazen vessels. In one of them, near the place where the enemy was forming his mine, the brazen vessels began to ring, from the blows of the mining tools which were at work. From this he found the direction in which they were endeavouring to penetrate.[18]

The fate of the miners is rather more specific than that of the Persian diggers: the Appolonians 'prepared vessels of boiling water and pitch, human dung, and heated sand, for the purpose of pouring on their heads. In the night they bored a great many holes, through which they suddenly poured the mixture, and destroyed those of the enemy that were engaged in this operation.'[19]

If we are to believe the accounts of later historians, acoustics could be a very useful science indeed in the classical period. Denis, ruler of Syracuse from 406 to 367 BCE, supposedly had such mastery of the practical construction of acoustic devices that he built a whole system of what we would call wave guides throughout his palace, so that he could spy on those who might be plotting against him (which was not an especially paranoid thing for him to do, given that he had shed a great deal of blood to reach his position of power). The wave guides terminated in listening points concealed behind statues, where Denis or his spies could lurk and listen.

Though something like this would have been technically possible, but there is no extant account of the story before the Middle Ages—and judging from the diagrams drawn then (see Fig. 6), the suspicion of the plotters, if there were any, might have been aroused by the architecture of the palace.

And then there was Alexander the Great, with his anecdotal and enormous megaphone, allegedly used to call his armies.

Whatever the truth of such stories, it seems certain that noise was already a widespread nuisance to many; a nuisance that could be acted on on a case-by-case basis by local legal representatives, as a piece of writing found in the ruins of Pompeii indicates: 'Macerior begs the magistrate to prevent the people from making a noise disturbing the good folks who are asleep.'[20]

FIGURE 6. Supposed acoustic devices in the Palace of Denis, tyrant of Syracuse.

From Athanasius Kircher, *Musurgia universalis sive ars magna consoni et dissoni in X. libros digesta* (2 vols; Rome: Ex typographia haeredum Francisci Corbelletti, anno Iubilaei 1650), ii, plate XVII; by kind permission of Bernard Becker Medical Library, Washington University School of Medicine.

By the first century BCE, public speaking was well established, and careful consideration of the noise issues related to it was well underway too. Lucretius, writing around 60 BCE, shows a clear understanding of the causes and problems produced by reverberation, still the most challenging aspect of the acoustic design of public spaces:

> When there is no long race for...voice to run from start to finish, each of the words...must necessarily be plainly heard...but if the intervening space is longer than it should be, the words...must be confused...perceived yet not distinguishable in meaning...so confused must be the voice when it arrives, so hampered...one voice... disperses suddenly into many voices...some scattered abroad without effect into the air: some dashed upon solid places and then thrown back...deluding with the image of a word...Therefore the whole place is filled with voices...all around boils and stirs with sound...[21]

To deal with the noise produced by workers, legal instruments were occasionally used: the council of the province of Sybaris, a Greek colony in the Aegean, ruled that potters, tinsmiths, and other tradesmen must live outside the city walls because of the noise they made. And, while they were at it, they banned roosters too. This is the first known noise ordinance and the first known example of officially organized noise zoning too, which has remained the most common way of managing most types of noise problem ever since. But such approaches remained—as far as we know—rare for many succeeding centuries.

This rarity may in part be due to the Romans, who felt that this sort of approach was a lot of fuss about nothing made by a decadent and over-refined people, whom they mocked with the tale of a Sybarite who could not sleep because there was a crumpled petal in his bed of roses. Seneca, a Roman writing in the first century CE while he was living above a gymnasium, summed up the official Roman line: noise was something to be got used to, and only the sort of thing that weak-willed people like Sybarites would allow themselves to be bothered by in the first place. This type of attitude has undermined the efforts of many anti-noise campaigners ever since.

For other Romans, however, putting this high-minded approach into practice was not so simple—and Seneca himself ends his essay 'On Noise' with the decision to move somewhere quieter. Horace, for instance, was clearly not greatly impressed by the idea that it was possible simply to adopt a superior attitude to city noise and carry on regardless, writing that: 'A contractor rushes along in hot haste with his mules and bearers. A huge machine is moving here a stone and there a beam. Sad funeral processions jostle with large wagons. Here a mad dog runs. There a mud bedaubed sow falls down. Go now and try to meditate on your melodious verses!'[22]

Juvenal had a similar problem:

> 'For what sleep is possible in a lodging in Rome? Only those with great wealth can sleep in the city. Here is the reason for the trouble. The movement of four-wheeled wagons through the narrow, winding streets, the clamorous outcries of the cattle drovers when brought to a standstill would be enough to deprive even General Drusus of his sleep.[23]

We don't know a great deal about Drusus, but presumably he was a no-nonsense kind of general. Or maybe a deaf one. Rather more succinctly, Martial said the noise of the streets at night made it sound as if the whole Roman army was marching through his bedroom. He also gives descriptions of typical street sounds of his time, including the shouts of beggars and the chink of bankers counting their coins in order to attract customers.

So, like the despised Sybarites before then, the Romans were sometimes reduced to legislation. In 44 BCE Julius Caesar ruled that, henceforward, 'no one shall drive a wagon along the streets of Rome or along those streets in the suburbs where there is continuous housing after sunrise or before the tenth hour of the day'.[24] Why he should restrict only *daytime* traffic is rather surprising, though perhaps Juvenal's comment that 'only those with great wealth can sleep in the city' may suggest a reason. We know that Pliny the Elder had a bedroom built with double walls so that he could sleep unaffected by the noises his slaves made, and it may well be that Caesar took similar precautions, so that, for him, it was the impact of noise on daytime activities that was the problem. Another Roman law prohibited coppersmiths from establishing their trade in any street where a professor lived, making a link between sensitivity to noise and intellectual pursuits that would be remade many times over the succeeding centuries.

The first great expert in practical acoustics whose name survives was Vitruvius, a Roman architect who worked in the first century CE and whose writings are full of the practical details that show his personal experience in the field of engineering as well as his firm grasp of the fundamental nature of sound itself. His account of its propagation, written nearly 2,000 years ago, is a clear and confident one and even uses the same example that is commonplace today:

> The voice is a flowing breath, made sensible to the organ of hearing by the movements it produces in the air. It is propagated in infinite numbers of circular zones, exactly as when a stone is thrown into a pool of standing water... Conformable to the very same law, the noise also generates circular motions, but with this distinction, that in water the circles remaining upon the surface, are propagated horizontally only, while the voice is propagated both horizontally and vertically.[25]

Speaking of the military engineer of his times, Vitruvius said of the role of harmony and discord in his setting up of war machines:

> Music assists him in the use of harmonic and mathematical proportion. It is, moreover, absolutely necessary in adjusting the force of the ballistae, catapult, and scorpions, in whose frames are holes for the passage of the homotona, which are strained by gut-ropes attached to windlasses worked by hand-spikes. Unless these ropes are equally

> extended, which only a nice ear can discover by their sound when struck, the bent arms of the engine do not give an equal impetus when disengaged, and the strings, therefore, not being in equal states of tension, prevent the direct flight of the weapon.[26]

However, Vitruvius cannot always have tested out the acoustic solutions he described with such confidence. He writes at length of the bronze 'sounding vessels' of the Greeks, which 'are fixed and arranged with a due regard to the laws of harmony and physics, their tones being fourths, fifths, and octaves; so that when the voice of the actor is in unison with the pitch of these instruments, its power is increased and mellowed by impinging thereon'.[27] Vitruvius goes on to advise that the best way to use such vessels is to deploy thirteen of them in cavities halfway up the sides of a room. Similar vases were indeed enthusiastically arranged in churches all over Europe for many centuries (the earliest known is in the eleventh century, the latest in the sixteenth). And indeed one might, *prima facie*, expect them to work—bottles do indeed resonate if you blow across the top of them, producing clear loud notes, and they do use the energy of your lungs to do so. In practice, however, the effect on the acoustics of a space is, pretty much, zero. Can no one have noticed? Or perhaps it was a case of the emperor's new clothes.

4

NEW NOISE

In the thirteenth century, Henry II introduced into English law the concept of nuisance, which survives to this day and which would ever after be the main route to legal redress for those who suffered from noise. But there are no contemporary accounts of its use for this purpose until a century later—even though gunpowder was rediscovered in Europe at about the same time and must have occasionally caused some very loud and alarming noises indeed, particularly as gunpowder-makers generally carried on their business in their homes.

The first record we have of an official noise complaint is from 1378. It was raised by the neighbours of an armourer and heard by the London Assize of Nuisance, to whom the armourer's landlord argued that men of any trade were free to work anywhere in London and to adapt their premises as they pleased. Unfortunately, the outcome of the case is unknown; and the records that survive from this period are so scattered and patchy that it is hard to judge how often noise complaints were heard by the Assize. The same patchiness affects the information we have about noises of all kinds, though there were a few locations that are commonly referred to as noisy ones. One such place was the priory of St Mary of Bethlehem, which, in 1375, became an institution for the mentally ill. Though it would not be nicknamed Bedlam for several decades, its reputation as a source of harrowing noises quickly spread.

Thanks to the inclusion of the study of music (which included many details of the Greek accounts of sound) in the 'quadrivium', the four standard subjects taught in universities, the educated few were by now fairly well versed in the theory of sound and noise. In fact, acoustical concepts were so familiar that Geoffrey Chaucer, in his *c.*1380 poem 'The House of Fame', could play with them, comparing sound sources with one of his favourite topics, the fart:

> Soun ys noght but eyr ybroken,
> and every speche that ys spoken
> lowd or privee, foul or fair,
> in his substaunce ys but air.[1]

Throughout Europe, church bells were unrivalled in their role of defining the boundaries of communities through sound, and there are records of complaints that bells cannot be heard clearly throughout their local parishes. In Catholic countries, bell-sounding was highly regulated: parish church bells were obliged to be silent until the most senior church in the region had been rung. Monasteries, no matter how large, had only a single bell to emphasize their humility, and in 1590 the Council of Toulouse ruled that monastery bells must not be so loud as to drown parish church bells. Monasteries in fact were much more interested in silence than bell-ringing: Franciscan monasteries had no bells at all, and there are records of medieval monasteries buying up adjacent lands to ensure a zone of quiet around them.

During the fourteenth century, however, the church's acoustic monopoly began to be challenged all over Europe by a new sound: that of the striking clock. Perhaps surprisingly, there is no evidence of complaints about the new invention, except for a few claims that the sounds of their bells soured beer.

From the fifteenth century there was another new sound source: the brass hunting horn. There is no reason to believe that these horns made a significant difference to the noise levels in the countryside, but, as a new component of the soundscape, their sounds are mentioned frequently from then on, in poetry and prose. To some extent, the fact that such noise sources are more reported does indicate that they were also more disturbing—more 'heard'—than more familiar sounds, even if they were no more powerful. So, it may be fair to say that the brass hunting horn really was a significant new noise source—if, of course, we can assume that the views of the tiny minority of people who could write were to some extent representative of the population.

A louder city

In the fifteenth century, the first references to London as a noisy place appear in the historical record, but it was not until the sixteenth that we begin to have accounts of what it actually sounded like. This early modern period has been called 'the age of the ear',[2] as it was a time when interactions between people were mostly face to face and the levels of literacy were very low. What this means in practice is that there are relatively few reports, but those that do exist are full of references to noises and other sounds.

Some of the issues of concern were very modern ones—traffic noise, for example, was a significant cause of complaint, with contemporary commentator John Stow reporting, just as his successors were to do every few years from then on, the problem caused by modern traffic in large volumes squeezing into streets not designed to accommodate them: 'the number of cars, days, carts and coaches, more than hath been accustomed, the streets and lanes being straitened, must needs be dangerous, as daily experience proveth.'[3] Stow's contemporary, the poet John Taylor, also refers to the danger and noise of the streets, 'where even the very earth quakes and trembles, the casements shatter, tatter and clatter'.[4]

However noisy it was by day, by night the city of London was a very much quieter place than it is now, with a curfew being in effect during which none but watchmen were allowed to walk the streets. The commencement of the restricted period (which, since as far back as the eleventh century, had begun at 8 p.m.[5]) was announced by the sound of the curfew bells. These were church bells, by now the clear sound of authority and control, underlining to the citizens just who was in charge of the world. A whole sequence of bells was heard every night, starting with that of St Mary-le-Bow in Cheapside. They would have been accompanied by the hubbub of people leaving taverns, apprentices going home from work, and the sounds of the closing and bolting of the city gates.

So London nightlife was an indoor, family matter (with no 'strangers' being permitted to stay in a private house for more than twenty-four hours). Far from people resenting these restrictions, some complained of bells being rung too late, with the effect of keeping apprentices too long at their work. If there were any sounds at night, they were loud and menacing, with the discovery of any criminal activity being announced with noise—the 'hue and cry', which it was every citizen's duty to raise if necessary.

By the end of the century, however, nights seem to have become noisier, to the extent that a new ordinance was announced in 1595, ruling: 'No man shall after the houre of nine at the Night, keepe any rule whereby any such suddaine out-cry be made in the still of the Night, as making any affray, or beating his Wife, or servant, or singing, or revelling in his house, to the Disturbaunce of his neighbours.' And, indicating that nights started and finished earlier then than now: 'No hammar man, as a Smith, a Pewterer, a Founder, and all Artificers making great sound, shall not[6] worke after the houre of nyne in the night, nor afore the houre of four in the Morninge.'[7]

During the day, the apprentices and other workers would often sing at their work, and this developed as the first small factories, which had appeared in the late sixteenth century, grew in number over succeeding decades. Song was particularly useful in keeping track of repetitive work like weaving—or, on the river, rowing, with chanting on its waves from the watermen who transferred goods along it and plied their trades on its banks. Different times and places on the river had their own distinctive chants.

Meanwhile, proletarian entertainment noise was often in conflict with more established forms. When the actor Richard Burbage proposed to reopen an old theatre in Blackfriars in 1596, there were complaints from local 'noblemen and gentlemen' that the noise of drums and trumpets would disturb church services; and not without reason: theatres of the time made up for what they lacked in visual special effects by using noise-making devices to the full, with fireworks and thunder machines being especially popular.

Noise and the law

One of the earliest legal cases that refers to noise and for which we have full details is from the 1560s, and relates to John Jeffrey, who let a room in his house to a schoolmaster. The room was next to Jeffrey's study, and Jeffrey found, not to anyone's great astonishment, that having schoolchildren next door is not very conducive to study. He took his case to court—who ruled that a school could be set up anywhere and its neighbours would just have to put up with it. Jeffrey studied elsewhere in future.

Zoning, that ever popular cure-all, now began to make its appearance in property leases, which occasionally specified the times or places at which certain noise-making activities were prohibited or permitted. The earliest such lease known today is that for a dancing school, dated 1610, which prevented dancing on the premises between 2 p.m. and 5 p.m. In 1617 the regulations of the Universities of Jena and Leipzig ruled that 'no noisemaking handworker' could work anywhere that was inhabited by 'Doktores'—since the noise would disturb their cogitations.[8] But such progress was scattered and slow: in the mid-1650s, the Cheltenham Assizes had to be stopped frequently because of street noise—yet there is no record of any attempt to take legal action against the noisemakers. For the moment at least even the movers and shakers seemed resigned to the inescapability of noise. Those who wanted quiet retreated to the peace of the country.

Meanwhile, the music of the spheres was still a well-known and widely accepted idea, but the accepted reason for its inaudibility had undergone a subtle shift—but a significant one in terms of what it tells us about the growing grasp of acoustics. While in antiquity the reason for the non-audibility was that the sounds were somehow 'too great' for our ears, now the reason they could not be heard was their shrillness: even Saturn, which has the widest orbit and hence the lowest pitch, made a noise that was far higher than the highest organ pipes or bird calls. One might even say that the concept of ultrasound had its birth with this 'explanation'.

The music of the spheres features in a contemporary satire called *Beware the Cat*. Written in 1584 by William Baldwin, it recounts the story of a man called Streamer, whose hearing undergoes a miraculous change so that suddenly he can hear all sounds (he achieves this by the remarkable technique of stuffing his ears with the fried ears of a cat, hare, fox, hedgehog, and kite, which nicely clears his ears of 'filthy rime'). Though Streamer can hear the planets very nicely as a result of his unprecedentedly clean ears, the terrestrial sounds that are audible to him are limited to those made within 100 miles (*c*.160 kilometres), due to the curvature of the Earth. (Though not correct, it is a nicely scientific touch).

Streamer's description of what he hears provides us with a unique glimpse into the sound world of the period. While his list of over forty noises includes animal cries of all kinds, from hogs to frogs, and such human sounds as scolding, counting coins, and weeping, absolutely no mechanical sounds or any others that we would usually think of as noises are included in it.

Other sources at the time mention particular sounds as problems, though there seem to be no complaints about overall noise levels. Correspondingly, noise was dealt with by laws that controlled specific sources, So, while there were no controls on the numbers of carts, a set of transport regulations issued in London in 1586 set fines for coaches or carts that creaked or squeaked through not being oiled.

Anecdotally at least, noise was far from a general concern: in 1598 a German visitor to England, Paul Hentzner, reported that the English were 'vastly fond of great noises', and a generation later, in 1627, the diarist John Evelyn, relating the sounds of St Paul's Walk, noted approvingly that so 'lowd a *Town*, is no where to be found in the whole world'.[9] At this time in fact, loud sounds were signals of celebration more often than danger, disturbance, or alarm, and few celebrations would be complete without bells, cannon, drums, or firecrackers.[10] From the late 16th century a sound which most would now find appalling was added to the London soundscape: the cries of baited bulls and bears. Special rings were constructed at Bankside for the 'sport' (one is shown in Fig. 7), which was not banned until 1835.

People who complain vociferously about noise—with no matter what justification—run the risk of being thought of as unreasonable, oversensitive, selfish, or just plain mad, and this is by no means a recent phenomenon. More than four centuries ago, just such a character played the central role in Ben Jonson's very urbane and up-to-the-minute play *Epicœne, or the Silent Woman*. Morose, 'a Gentleman that loves not noise', was portrayed throughout as a figure of fun, the butt of everyone's jokes. In portraying how Morose copes with the noises that so disturb him, the play offers a glimpse of the ways in which people thought they could be dealt with; as one character explains '[Morose] was wont to go out of Town every Saturday at ten a Clock, or on Holy-day Eves. But now, by reason of the sickness, the perpetuity of ringing has made him devise a Room, with double Walls,[11] and treble Cielings; the Windows close shut and calk'd . . . '.[12]

The 'sickness' mentioned here is the bubonic plague, which was to be a frequent visitant to London throughout much of the century. Though its most deadly and famous outbreak was in 1665, 1609 (when *Epicœne* was written) was an especially bad year, with a mortality of 4,240. This figure was more than 2,000 higher than any other year from 1604 to 1623. As all of London's 114 churches tolled their bells long and frequently for the dead, the sound of London at the time must have been as uniform as it was doleful.

FIGURE 7. London in 1560, including a bull-baiting arena in the foreground. Private Collection/Bridgeman Art Library.

We also have a non-fictional account of the noise of London at this time. Written about two years before *Epicœne,* Thomas Dekker's *The Seven Deadly Sinnes of London* ascribes the coming of the plague to the combined effects of fraud, lying, sloth, deeds of darkness, cheating, cruelty, and, oddly enough, 'changes of fashion'. Dekker describes how the city of 1606 sounded:

> in every street, carts and Coaches make such a thundering as if the world ranne upon wheeles: at everie corner, men, women, and children meete in such shoales, that postes are sette up of purpose to strengthen the houses, least with jostling one another they should shoulder them downe. Besides, hammers are beating in one place. Tubs hooping in another. Pots clinking in a third, water-tankards running at tilt in a fourth...[13]

Meanwhile, in *Epicœne*, Morose is also tormented by a noisy band of interlopers who invade his home and shatter his peace. Such groups, called charivari,[14] were well known in the real world at the time as the latest form of what came to be called 'rough music': the use of noise by groups representing local opinion, to torment others (see Fig. 8). While in Morose's case the annoying group

FIGURE 8. William Hogarth's 1822 image of a skimmington, a type of charivari that usually included dummies representing the targets of the crowd. <http://www.CartoonStock.com>.

arrives ostensibly to 'celebrate' his marriage (which is, to no one in the audience's great surprise, to a very noisy woman indeed). The idea of a charivari seems to have struck many rural communities in different parts of Europe at about the same time, and they were sometimes used to coerce unmarried couples to wed. In other cases they were expressions of disapproval of anyone who transgressed conventions: married couples of very different ages, single mothers, adulterers, even marriages that were felt to take place too soon after widowhood. Though violence was sometimes involved, the main weapon of the charivari was simply noise. One, which took place in 1618, was led by a drummer and included several hundred participants. Their victim on that occasion was a supposed cuckold and his wife, and the gang terrorized them both by a violent attack and by the noise they made: 'the gunners shot off their pieces, pipes and horns were sounded, together with lowbells and other smaller bells which the company had amongst them.'[15]

5

A NEW SCIENCE OF SOUND

By the early seventeenth century sudden flowerings and rapid progressions characterized many areas of science—but the development of the science of sound was languishing. As Francis Bacon, the great agenda-setter of science in the period, noted: 'the nature of sound hath in some sort been inquired, as far as concerneth music, but the nature of sound in general hath been superficially observed. It is one of the subtlest pieces of nature.'[1] But he was confident that this state of affairs would not persist for much longer: in his visionary novel *The New Atlantis*, published in 1624 (and occasionally regarded as the first science-fiction novel), he put forward his idea of a scientifically based utopia. Among other things he anticipated 'sound houses' (see Fig. 9), where

> we practise and demonstrate all sounds, and their generation. We have harmonies which you have not, of quarter-sounds, and lesser slides of sounds. Divers instruments of music likewise to you unknown, some sweeter than any you have; together with bells and rings that are dainty and sweet. We present small sounds as great and deep; likewise great sounds extenuate and sharp; we make divers trembling and warblings of sounds, which in their original are entire. We represent and imitate all articulate sounds and letters, and the voices and notes of beasts and birds. We have certain helps which, set to the ear, do further the hearing greatly. We have also divers strange and artificial echoes, reflecting the voice many times and, as it were,

FIGURE 9. Illustration of one of Bacon's sound houses, together with a range of acoustic devices. Not to mention a giant strawberry.
From Francis Bacon, *The New Atlantis* (1620).

tossing it; and some that give back the voice louder than it came; some shriller, and some deeper; yea, some rendering the voice differing in the letters or articulate sound from that they receive. We have also means to convey sounds in trunks and pipes, in strange lines and distances.[2]

Bacon is also the first person to describe the phenomenon of temporary threshold shift, in which the ears become muffled after exposure to a loud sound. The cause is a defensive reduction in blood supply to the hair cells of the basilar membrane. In his case the cause was a hunting horn:

A very great sound, near at hand, hath strucken many deaf; and at the instant they have found, as it were, the breaking of a skin of parchment in their ear: and myself standing near one that lured 3 loud and shrill, had suddenly an offence, as if somewhat had broken or been

> dislocated in my ear; and immediately after a loud ringing (not an ordinary singing or hissing, but far louder and differing) so as I feared some deafness. But after some half quarter of an hour it vanished.[3]

Ever an enthusiastic collector of facts, Bacon goes on to collate other accounts of noise and its effects, including such handy tips as that 'discharge of artillery is injurious to lobsters'.[4]

The inaudible music of the spheres enjoyed a new surge in popularity, thanks to the work of Johannes Kepler. His *On the Harmony of the World* (1619) is a mix of science, superstition, and speculation.[5] Most of the massive tome consists of a meandering series of speculations that sound very strange indeed to a modern reader. It does contain a *bona fide* scientific breakthrough, but to reach it one first has to wade through Kepler's theory that the planets can be identified with singers: Jupiter is a bass, Mars a tenor, Earth a contralto, and so on. Despite being the man who overturned centuries of stubborn belief that the planets must move in perfect circles by showing that in fact they travel in ellipses, Kepler is still influenced by the idea of the superiority of the circle: he uses the departures of the orbits of the planet from the 'perfection' of circularity to modify the note each 'sings', by turning it into a chord. The Earth chord is the combination of notes that, if sung, would be Mi, Fa, and Mi. Mi he considers to stand for Misery and Fa for Famine, explaining our gloomy conditions here on Earth: the idea that Earth was a terribly decayed, disreputable kind of world, and not a patch on the other planets, the Sun and so on, was commonplace among thinkers from the Greeks up to, but not including, Galileo.

Galilean acoustics

As Bacon lamented, for centuries the progress of acoustics achieved very little in terms of new scientific knowledge. The conclusions of the Ancient Greeks were simply applied to new situations, along with a little modernizing here and there, but no one carried out any new research until the time of Galileo Galilei. Galileo challenged and changed many areas of scientific knowledge, formulated entirely new ones, and began to put scientific research on a firm mathematical footing. In addition to his unparalleled achievements in investigating the nature of motion of many kinds, he also made the first scientific use of the telescope, using it almost literally to push back the boundaries of the Universe, and to provide good evidence that Copernicus was right in his conjecture that the planets revolved around the Sun. In addition, he investigated afresh the phenomena of acoustics—and, like Pythagoras before him, he began his investigations with the nature of harmony and of dissonance.

But in this area Galileo was not the first. It was his father Vincenzo who made the first new fundamental contribution to the science of dissonance for many centuries.

Since antiquity, the core and foundation of the theory of music had been a mathematical account of harmony and discord: what, scholars asked, are the mathematical ratios of the lengths of strings that produce consonances when sounded in sequence, or together? And how is an octave best produced? It had always been thought that the answer to the latter question was very simple: an octave interval would be produced by strumming two strings that were identical other than their tensions, if one of these tensions was twice that of the other. This could be most simply achieved by suspending a pair of weights from them, one twice as massive as the other.

This idea had been accepted for centuries before Vincenzo showed its falsity, and gave the required ratio of tensions as not 2:1 but 4:1. Similarly, he proved that, while one can achieve a perfect fifth with a pair of strings bearing the *length* ratio 3:2, if one strums a pair of strings whose *tensions* are in this ratio the result is not harmonious at all, but really rather horrible. Instead, what one needs are weights in the ratio of 9:4. In other words, while the ratio of an interval is proportional to string lengths, it is proportional to the *square roots* of the tensions applied. This is the earliest scientific law in which square roots play a part—and therefore the world's first non-linear physical formula.

Some of Vincenzo's other important contributions to musical theory involve the treatment of dissonance. In particular, he introduced a concept of 'passing' dissonance, applicable to notes sounded in quick succession and distinct from on-the-beat dissonance, which he called 'essential' dissonance. Vincenzo also defined rules for the use of tension in music by a preliminary leap away from, followed by a return to, a consonant note.

Vincenzo did much of his work on acoustics in the 1580s, when his son Galileo was living at home giving private lessons in mathematics. With this heritage of dissonance ringing in his ears, Galileo was in an ideal position to push forward the science of acoustics further, and so he did. His grasp of harmony and dissonance is assured, masterly, and completely correct, and he was also the first to pin down the physical relationship between pitch and frequency. It is one of the charms of Galileo's science that he explains it, not through dry essays or reports, but in the form of a playlike dialogue—in this case between Sagredo, a breathlessly excited man keen to learn all the new world of science can teach him, and Salviati, a stand-in for Galileo himself, who consequently knows all the answers. The third character in the dialogues is Simplicio. The trio had previously appeared in Galileo's even more celebrated—and reviled—*Dialogue on the Two Chief Systems of the World*, which Galileo had written with the agreement of the Pope to propound and compare the Copernican and Ptolemaic theories of the Solar System.[6]

Unfortunately, Simplicio is a bit of a dunce, and a mouthpiece for the Pope's view too, so it is no surprise that the Pope decided he was based on none other than himself. This played no small part in the book being banned and Galileo being sentenced to house arrest for the rest of his life. It was from this state of house arrest that *Two New Sciences* was written. Rather courageously Galileo included Simplicio again—as slow-witted as ever. As he was forbidden to publish anything in Italy, *Two New Sciences* appeared instead in the Netherlands (in 1638).[7] The sciences of the title are those of the strength of materials and of dynamics.

Here, Salviati explains the observations that led him to his discovery of the relation between frequency and pitch:

> That the undulations of the medium are widely dispersed about the sounding body is evinced by the fact that a glass of water may be made to emit a tone merely by the friction of the finger-tip upon the rim of the glass; for in this water is produced a series of regular waves. The same phenomenon is observed to better advantage by fixing the base of the goblet upon the bottom of a rather large vessel of water filled nearly to the edge of the goblet; for if, as before, we sound the glass by friction of the finger, we shall see ripples spreading with the utmost regularity and with high speed to large distances about the glass. I have often remarked, in thus sounding a rather large glass nearly full of water, that at first the waves are spaced with great uniformity. When, as sometimes happens, the tone of the glass jumps an octave higher, I have noted that at this moment each of the aforesaid waves divides into two; a phenomenon which shows clearly that the ratio involved in the octave is two.[8]

Galileo is also the first to attempt properly to explain why it is that dissonances sound nasty:

> I assert that the ratio of a musical interval is not immediately determined either by the length, size, or tension of the strings but rather by the ratio of their frequencies, that is, by the number of pulses of air waves which strike the tympanum of the ear, causing it also to vibrate with the same frequency. This fact established, we may possibly explain why certain pairs of notes, differing in pitch, produce a pleasing sensation, others a less pleasant effect, and still others a disagreeable sensation. Such an explanation would be tantamount to an explanation of the more or less perfect consonances and of dissonances. The unpleasant sensation produced by the latter arises, I think, from the discordant vibrations of two different tones which strike the ear out of time. Especially harsh is the dissonance between

> notes whose frequencies are incommensurable; such a case occurs when one has two strings in unison and sounds one of them open, together with a part of the other which bears the same ratio to its whole length as the side of a square bears to the diagonal; this yields a dissonance similar to the augmented fourth or diminished fifth... Thus the effect... is to produce such a tickling of the eardrum that, allaying the sweetness by a mixture of tartness, it seems at once to gently kiss and to bite.[9]

Galileo then goes on to draw an analogy with the motion of multiple pendulums of different lengths: if those length are related by integers, the pendulum bobs will make pleasing patterns of motion in which they frequently swing together—equivalent to harmonious sounds. If the lengths are not so related, they will present a disorderly mix of motions, just like dissonant sounds. And indeed to this day the motion of pendulums is referred to as 'simple harmonic', and the shape that each traces out (for instance, on a strip of paper that is allowed to unroll steadily under it) is that of a sine wave—the same shape that characterizes the pressure variation of a steady sound wave (see Fig. 10).

It is not surprising that Galileo could go no further in his account—even today the mechanisms involved are uncertain, and the fact that notions of dissonance are neither quite objective nor quite subjective adds to the puzzle: while everyone everywhere seems always to have agreed that octaves and perfect fifths are consonant, there is wide variation on which other combinations are judged nice or nasty—though there is agreement that the worst

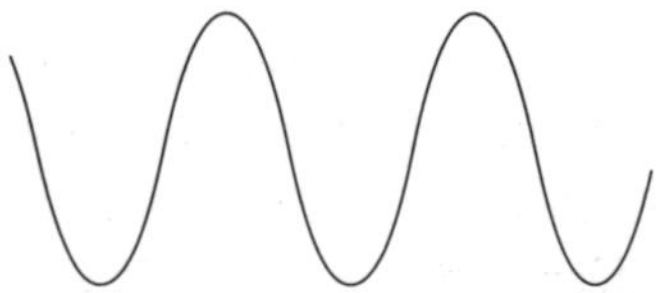

FIGURE 10. Sine wave.

dissonance is the half-octave, or tritone. As Galileo noted, the key characteristic is that, while octaves and fifths can be expressed in the very simple whole-number ratios 2:1 and 3:2, the tritone is not a ratio of whole numbers as all, but is instead given by $\sqrt{2}$:1. Pythagoras would no doubt have been very pleased to hear all this (if rather perturbed to find that the secret of irrational numbers was well and truly out). Sounds with no harmonies at all, like those of nails scraped across blackboards, also seem to be universally loathed—but only by humans. Monkeys don't seem to mind them at all. A full theory of dissonance has to explain not only all this but also why it is that children are so much less tolerant of dissonance in music than adults—as anyone who has had to listen to a CD of songs for children is only too aware. But no further development in this area was to take place for the next two centuries.

Technologies of noise

Meanwhile, there was a new spirit of scientific enquiry in the air, and a number of scientists began to work extensively on the science of noise. Pre-eminent among these was Athanasius Kircher, whose 1673 book *Phonurgia Nova sive Conjugium Mechanico-Physicum Artis & Naturae Paranymta Phonosophia Concinnatum* was the first to deal exclusively with acoustics.[10] Kircher occasionally lapsed into an Aristotelian approach, as in his suggestion that the music of the spheres does indeed exist but is inaudible because our ear passages are too narrow to admit it—a view that shows a lack of understanding of what the 'sizes' of sound waves really amount to. However, Kircher was a practical experimenter as well as a theoretician, and he invented a plethora of acoustic devices, like the one shown in Fig. 11.

Kircher also built at least one of his inventions: the megaphone.[11] Though various Greek and Roman antecedents had been mentioned, there is no clear evidence that any of them actually existed. Kircher's version is the first of which we are certain—and it was an impressively noisy affair: 5 metres long, made of iron plate, and 60

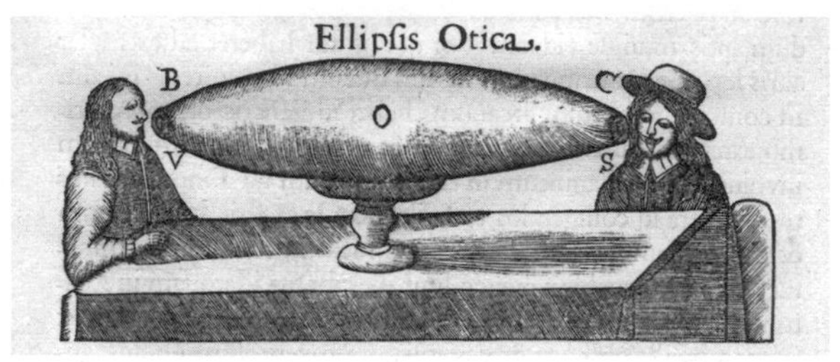

FIGURE 11. Illustrations of a 'hearing lens'—rather unwieldy hearing aids—from Kircher's *Phonurgia nova.*

Medicine: Illustration of the Ellipsis Otica, from Athanasius Kircher, *Phonurgia nova sive conjugium mechanico-physicum artis & naturae paranympha phonosophia concinnatum* (Campidonae Kempten: Per Rudolphum Dreherr, 1673).

centimetres wide at the business end. It opened into the wall to his garden in Rome and turned out to be ideal for bellowing at distant servants and for conversing with his gardener without the need actually to leave his room and meet him. So excited was Kircher by the possibilities of his invention that he extended its use to startling guests with apparently speaking statues—and to eavesdropping on them too. He even proposed to broadcast the sound of indoor music to dancers outside, though how exactly the musicians would be able to direct their efforts down the end of the megaphone was left unclear.

An obvious shortcoming of this device was that all the excitement was confined to his house and garden—so his next invention was a 'portable' (if only by pairs of burly servants) version, 3 metres long and 1 metre in diameter. No doubt to the delight of the local church, its first public use was to summon 2,200 of the faithful to a special service. There are unfortunately no records of the reactions of the participants.

Meanwhile, across the Channel, Sir Samuel Morland was having a similarly loud and entertaining time with his own version of the

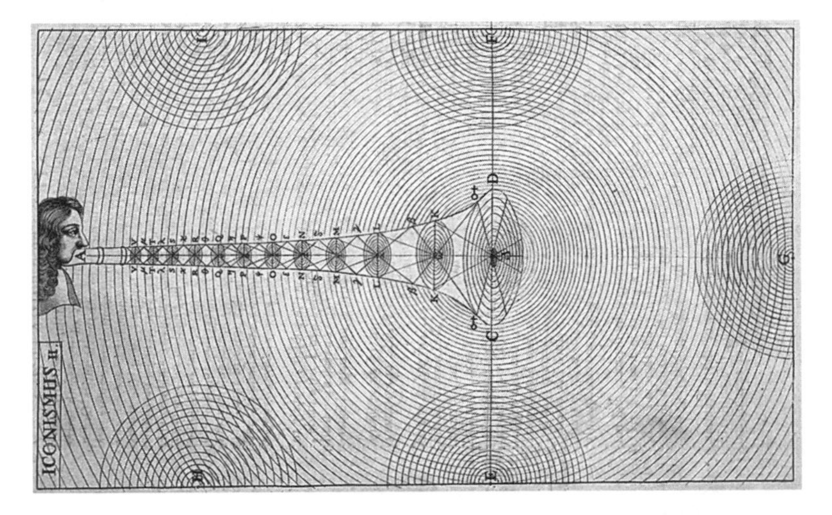

FIGURE 12. Morland's diagram of the sound field of his Tuba-Stentoro-Phonica.

From S. Morland, *Tuba Stentoro-Phonica, An Instrument of Excellent Use, As well at Sea, as at Land; Invented and Variously Experimented in the Year 1670. And Humbly Presented To the Kings Most Excellent Majesty Charles II. in the Year 1671* (London, 1672), 8 (STORE 131:38); by kind permission of the University of Cambridge, Whipple Library.

megaphone. The prototype was, in Kircher's terms, rather minuscule, being only 80 centimetres long with a mouth 28 centimetres wide, but it was shortly followed by a monster version, 6.4 metres in length (see Fig. 12).

The invention was 'humbly presented to the Kings Most Excellent Majesty Charles II, in the year 1671',[12] and, just in case Charles was rather at a loss as to what to do with it, Morland wasn't short of ideas. According to its fond inventor, the possibilities of the invention were staggering. Among the 'manifold uses' of the impressively named Tuba-Stentoro-Phonica[13] were:

> In case a Town or City be Besieged, and so close girt about, that there can be no message sent in ... And so on the contrary, may the Besiegers make as good use of this instrument to Threaten and discourage the Besieged ... In case of great Fires, where usually all

> people are in a hurry...In case a number of Thieves and Robbers attaque a House that is lonely, and far from Neighbours, by such an Instrument as this, may all the Dwellers around about, within the compass of a Mile or more, be immediately informed, upon whose House such an attaque is made, the number of Thieves or Robbers, how armed and equipped, what manner of persons, with the colour and fashion of their Habits, and by what way they have made their escape...or which way to pursue them.[14]

Morland used his invention for scientific research as well as fun. He was the first to investigate the directivity of noise sources, by mapping out the variation in loudness around the horn, even sticking his head inside as far as he could. The only measuring instruments he had were his ears, but his conclusion that the sound was greatest at the centre of the mouth was quite correct.

Newtonian noise

Despite this newfound interest in noise and noise-making, little further progress was made in the scientific understanding of any aspect of acoustics until Isaac Newton, the giant of seventeenth-century science, turned his attention to it. The contributions he made to the science of sound were as influential as those he made to many other areas of physics, and he was the first to apply a technique that would be the mainstay of most aspects of acoustics from then on: the construction of a model, built from pure mathematics, that could make testable predictions to corroborate a theory's accuracy as a description of the physical world.

For seventeenth-century scientists the construction of such models was so challenging that they could be applied only to one of sound's most basic and easily grasped properties, its speed. Where Newton went, other scientists, for many decades after, were prone to follow—usually with great justification. So it is a pity that Newton's approach to this particular topic was, to say the least, a bit odd.

Accepting the account that had been prevalent since antiquity, that sound is a progression of pressure pulses through a medium, Newton in his monumental and truly revolutionary work *Philosophiæ Naturalis Principia Mathematica* studied the nature of such a progression from first principles.[15] To do this, he used calculus, a branch of mathematics that he had invented (and that he called fluxions, since it is designed to deal with changing or 'flowing' quantities). But, with a regard for secrecy that bordered on paranoia, he kept quiet about this and instead used tortuous geometrical demonstrations to prove his conclusions, making it even more difficult for others to spot any weaknesses in them.

After one has struggled through the geometrical obfuscation, a correct account of sound waves is revealed: particles of a fluid move backwards and forwards in obedience to the same law that governs the motion of a pendulum—simple harmonic motion. That is, the particles oscillate with a constantly varying speed, which accelerates during their approach to the midpoint and then decelerates once that point is passed until they stop and begin to accelerate back again.

When an object moves, it sets up harmonic motions in particles adjacent to its surface, and these motions are passed to neighbouring particles and then to further ones, such that a pulse of pressure travels through the medium; a sequence of such pulses forms a sound wave. Newton showed that 'the velocities of pulses propagated in an elastic fluid are in a ratio compounded of the subduplicate ratio of the elastic force directly, and the subduplicate ratio of the density inversely'[16]—or, as we would say, the speed of sound is proportional to the square root of the elasticity over the square root of the density of the medium. Newton correctly demonstrated too that the speed of sound is independent of its intensity (not at all obvious in its day), and also pointed out the relation between velocity, frequency and wavelength: $v = f\lambda$.

Throughout the *Principia*, a key element of Newton's approach—which became a guiding principle of science thereafter—was to 'contrive no hypotheses'—that is, not to make up theories that are not well grounded on the facts available. Unfortunately, Newton was dealing, even in this apparently simple situation, with topics with which the concepts of his time were too primitive to grapple. The underlying problem with his analysis is his assumption that the elasticity of the medium through which sound travels is proportional to its 'condensation'. It turns out that this is equivalent to the concept, not formulated clearly until long after Newton's time, that the expansions and contractions of the medium are isothermal—that is, they take place without change of temperature. This is incorrect, and in fact the temperature oscillates just as the particles' positions do, rendering Newton's whole approach fatally flawed.

Newton's calculations led him to predict a speed of 968 feet (295 metres) per second—a figure lower than the actual value by about 16 per cent. However, contemporary measurements of the speed were primitive, and Newton's result was just within the range of measured values he found, of 866 and 1,272 feet (264 and 388 metres) per second.

The first edition of the *Principia*, including Newton's work on sound, was published in 1687. In 1698 and 1708 new measurements of sound speed narrowed the margin for error considerably, and revealed that Newton's value was definitely incorrect. A value for the speed of 1,142 feet (348 metres) per second was accepted (and is correct, for typical temperatures and pressures).

Rather breathtakingly, in the 1713 edition of the *Principia*, the value of sound speed that Newton derived had changed from 968 to 1,142 feet per second. Even more so is the way in which this much improved result was achieved. First, Newton redid his calculations with slightly different values for the medium, giving him a speed of sound of 979 feet (298 metres) per second. Then he decided he really ought to make some allowance for the 'crassitude' of the air particles—109 feet (33 metres) per second seemed suitable (though we are uninformed as to why this correction should be necessary or where the number comes from—or, for that matter, what 'crassitude'[17] might be). This now gave 1,088 feet (332 metres) per second. Close, but not good enough for Newton. He decided that it would really be very wise indeed—and not a bit post-hoc—to multiply the figure by 21/20, to compensate for 'the vapours in the air being of another spring',[18] which gives the required value of 1,142 feet per second (348 metres per second) very nicely. The correct answer, and hardly any hypotheses contrived at all.

What are we to make of Newton's behaviour here? His reaction, I suspect, would be to say that he wished he had never published his theory of sound in the first place, just as he did when his theory of light came in for criticism, saying: 'I see I have made myself a slave to philosophy... I will resolutely bid adieu to it eternally, excepting what I do for my private satisfaction, or leave to come out after me; for I see a man must either resolve to put out nothing new, or to become a slave to defend it.'[19] And, indeed, he published his theory of the speed of sound along with the rest of the *Principia* only at the urging of his friend and colleague Edmond Halley. Perhaps it was his view that one could simply not be seen to be wrong in print—combined with the suspicion and secrecy so central to his nature—that caused this hesitation.[20]

Whatever their cause, Newton's actions caused problems. If he had been less well respected, his reanalysis of the speed of sound might have been politely ignored or vehemently corrected, but such was the awe in which he was viewed after the publication of the *Principia* that to doubt his statements felt almost like blasphemy to scientists—especially English ones. So his approach stuck for centuries. Since the measured speed of sound stubbornly remained around 1,142 feet per second, many great scientists over the next few decades wasted their time fiddling with Newton's formula without altering his assumptions. One such was Leonhard Euler—one of the greatest mathematicians of his or any age—who simply inserted a $4/\pi$ factor into his version of Newton's formula, with nothing to justify it except a nice answer.[21]

It was not until over a century later, in 1816, that the problem was correctly analysed, by Pierre Simon Laplace. Laplace made just one key contribution to acoustics: that sound waves are not isothermal but adiabatic—that is to say, in contradiction of Newton, that the temperatures of the particles involved *do* change as a sound wave passes, but there is no overall loss or gain of heat. With this change, the theory became clear, and the predictions became correct. Even today, Laplace's version of Newton's theory holds true in most circumstances.[22]

Newton's contemporary and sometime rival Robert Hooke had a more empirical approach to the study of sound, and in fact seems to have anticipated many modern approaches to noise. He suggested in his posthumously published *A Curious Dissertation concerning the Causes of the Power & Effects of Musick* that, since babies respond to music, an awareness of harmonious sound waves must precede an awareness of language. Of noise, Hooke suggested that it 'is displeasing because the ear cannot keep up with the constant change of tuning required'.[23]

Hooke also loved the idea of extending the power of the ear—but not for the purposes of crime prevention or assistance to the hard of hearing, as Morland and Kircher had suggested: he wanted to use sound like one of the microscopes he was so fond of, to study

> the possibility of discovering the internal motions and actions of bodies by the sound they make... Who knows, I say, but that it may be possible to discover the motions of the internal parts of bodies, whether animal, vegetable or mineral, by the sound they make; that one may discover the works performed in the several offices and shops of man's body and thereby discover what instrument or engine is out of order, what works are going on at several times and lie still at others and the like... I have been able to hear very plainly the beating of a man's heart- and it is common to hear the motion of wind to and fro in the guts and other small vessels...[24]

No doubt Hooke would have been most impressed if he had known that today the sound of turbulence in arteries is used as a way of detecting their narrowing, and that equivalent systems in crops like tomatoes use the sounds of fluid flow not only to monitor health but automatically to activate watering systems (and the ripeness of the tomatoes can be assessed by measuring the speed of sound through them, too).

Hooke was also the first to demonstrate unambiguously that frequency is related to pitch, through the simple but effective means of holding a piece of card against a rotating toothed wheel. As the frequency of rotation of the wheel increased, so did the pitch of the sound produced. Hooke's contemporary Robert Boyle was the first person successfully to carry out the well-known demonstration that a ringing bell ceases if the air around it is removed. Key to his success was the effectiveness of the air pump that he built—it was the poor performance of earlier air pumps that had led some other experimenters to conclude that sound, like light, can travel through a vacuum. It is rather unfortunate that this, perhaps the best known of all acoustical demonstrations, does not actually prove what it is often supposed to. The reduction in loudness as the air leaves the vessel is due mainly to the difference in densities between the air inside and outside the vessel: unless the vacuum is very hard (harder than was obtainable in the seventeenth century), plenty of acoustic energy makes its way from the bell to the vessel walls, but is then reflected back again. This is known as impedance mismatching. Impedance mismatching also explains why elderly campers who claim that newfangled metal tent pegs are not a patch on good old wooden ones are in fact talking nonsense. Wooden tent pegs are much harder to hammer into the ground, owing to the same law of acoustics that silences the bell. The acoustic impedance is the product of the density of a material and the speed of sound in it, and this amount is much more similar in wood-and-earth than in steel-and-earth—so hammering at a wooden tent peg results in most of your effort passing through the peg into the ground, with the result that the peg sits there laughing at you and hardly moving. In the case of the metal peg, the impedance mismatch between it and the earth is so large that the energy is mostly confined to the peg and is expended in forcing it to go neatly and smoothly into the ground. Similar effects explain why impedance mismatching is key to sound insulation.

6

NOISE IN THE EIGHTEENTH CENTURY

In the first history of acoustics, *Anecdotal History of the Science of Sound to the Beginning of the Twentieth Century* (1935), its author, Dayton Clarence Miller, says rather despairingly that 'a careful search fails to reveal any major contributions to the science of sound which arose in the 18th century',[1] and indeed there were very few, though, in 1743, Abbé J. A. Nollet did conduct a series of experiments to settle a dispute about whether sounds could travel through water. By the simple expedient of submerging his head in a lake, he discovered that he could indeed hear a bell, gunshot, shouting, and whistling. A slightly less obvious observation was that a bell struck underwater could be heard more easily through the water than through the air above it. We know now that the ear is considerably less sensitive in water than in air, but in this case this is more than offset by the fact that that sound is much more effectively transmitted by water.[2]

While the science of noise was making little headway, noise itself was developing rapidly. By now, the idea that noise was vulgar was well established, along with all sorts of other codes of social conduct and etiquette that acted as an infallible guide to allow the trained observer to spot instantly who was sophisticated and elegant and who was common and rowdy. Meanwhile, in Swiss cities, particularly in Bern, by-laws were being introduced specifically against noise: the first, passed in 1628, was 'against singing and shouting in streets or houses on festival days',[3] and additional such laws were introduced every few years from then on, peaking in the 1910s, when six were passed. In the UK meanwhile, while no specific anti-noise legislation was to appear for many years, a number of laws were introduced that, though intended for other purposes, acted to reduce noise. There were licensing laws that limited the times of the public consumption of alcohol and the rowdiness that arose from it. Church regulations required congregations to be quiet, so that sermons could be heard properly, and the work of apprentices was required to cease by 9 p.m. at the latest—this last law was introduced because of the risks associated with the use of candles. The effects were to restrict noise-making somewhat to certain times of day, zoning time, rather than space, in these cases.

At the time, a famous source of noise was the numerous fairs that were held regularly in many countries. In England, perhaps the most famous was the Bartholomew Fair, so well known that Ben Jonson wrote a play about it in 1719. The purpose of *Bartholomew Fair* was to satirize Puritans, fortune-hunters, country bumpkins, and inept representatives of the justice system—but in so doing Jonson also gives us an insight into the sorts of noise that might be heard there. Here is Leatherhead, a 'Hobbi-Horse'[4] seller, proffering his noisy wares to one of the visitors, Zeal-of-the-Land Busy:

> LEATHERHEAD: What do you lack, Gentlemen? What is't you buy? Rattles, Drums, Babies—
>
> BUSY: Peace, with thy Apocryphal Wares, thou profane Publican: thy Bells, thy Dragons, and thy Tobies Dogs. Thy Hobbi-horse is an Idol, a very Idol, a fierce and rank Idol: and thou, the Nebuchadnezzar, the proud Nebuchadnezzar of the Fair, that set'st it up, for Children to fall down to, and worship.
>
> LEATHERHEAD: Cry you mercy, Sir; will you buy a Fiddle to fill up your noise?
>
> . . .
>
> LEATHERHEAD: Or what do you say to a Drum, Sir?

BUSY: It is the broken Belly of the Beast, and thy Bellows there are his Lungs, and these Pipes are his Throat, those Feathers are of his Tail, and thy Rattles the gnashing of his Teeth.

...

LEATHERHEAD: Sir, if you be not quiet the quicklier, I'll ha' you clapp'd fairly by the Heels, for disturbing the Fair.

BUSY: The Sin of the Fair provokes me, I cannot be silent.

PUR [DAME PURECRAFT]: Good brother Zeal!

LEATHERHEAD: Sir, I'll make you silent, believe it.

....

[Leatherhead enters with Officers]

LEATHERHEAD: Here he is, pray you lay hold on his Zeal; we cannot sell a Whistle for him in tune. Stop his noise first.

BUSY: Thou canst not; 'tis a sanctified noise...[5]

The reference to sanctified noise here is an interesting one: at the time it was the sound of church bells that was 'sanctified' in the sense of being immune from criticism as representing the most powerful authority in the land—as is clear from the absence of complaints about them.

Meanwhile, in many parts of London, street criers were increasingly referred to as sources of noise—and their number was reaching the point of diminishing returns in some areas, with so many simultaneous street cries that people would often be confused as to who was selling what.

At the same time, there was a steadily rising number of urban professionals, who wished to distance themselves from such tradespeople, and also from the rowdy entertainments of the working classes. For centuries, perhaps millennia, noise had been regarded by the educated as unsophisticated and vulgar, but this view may have become much more widespread and outspoken at this time through the rise of this new class.

As its members were literate, we have plentiful records of their views—though, of course, these records cannot be regarded as anything like a balanced account. According to them, both street music and street criers were thought to be a general nuisance (see Fig. 13). Jonathan Swift, for example, complained about cabbage sellers: 'here is a restless dog crying cabbages and savoys, plagues me mightily every morning about this time. He is at it now. I wish his largest cabbage was sticking in his throat.'[6] Samuel Butler meanwhile claimed that those who 'have but a little wit are commonly like those that cry things in the streets, who if they have but a Groatsworth of Rotten or stinking stiff, every body that comes nigh shall be sure to heare of it, while those that drive a rich noble Trade make no Noyse of it'.[7]

FIGURE 13. Study of a street crier, with a donkey, William Henry Bunbury, late eighteenth century.
Private Collection/Christie's Images/Bridgeman Art Library.

It is not until 1713 that we find any evidence of interest on the part of medical professionals in the impact of noise on workers—specifically, on Venetian coppersmiths. It is found in the first significant account of occupational disease, *De Morbis Artificum* ('Diseases of Workers'), by Bernadini Ramazzini, an Italian physician, who says that copper hammerers 'have their ears so injured by that perpetual din . . . that workers of this class become hard of hearing and, if they grow old in this work, completely deaf'.[8]

It was to be many years yet before such information was to be acted on, but, at about the same time, a thin trickle of complaints about noise begins to appear in public records, though, as no general laws against noise had been passed anywhere, the opportunities for legal redress were limited. What was possible had in fact been so for some time, but it was only now that anything was actually done.

And what was possible? As William Blackstone, a legal writer, put it in his 1770 *Commentaries on the Laws of England*: 'Nusance [*sic*], nocumentum, or annoyance, signifies any thing that worketh hurt, inconvenience or damage,' and an annoyance was actionable if it interfered with the ability of another to use or enjoy his land (a reminder that law in those days was very much concerned with the protection of property), which sounds very promising as a weapon to use against noise. And indeed, among the apparently miscellaneous list of nuisances included by Blackstone, there were things like 'using a speaking trumpet' (much to Morland's disgust no doubt) and throwing fireworks. A distinction was drawn between public nuisances—which had the potential to annoy anyone in the vicinity—and private nuisances, which affected the land or property of individuals or small groups. To avoid excessive litigation, in the case of public nuisances one single representative case was selected and heard in court in the name of the king or queen. Private nuisances, on the other hand, could be dealt with under the law only through individual private legal actions.

The novelty of taking official action to deal with noise problems is suggested by the fact that the response of the Court of London Aldermen when petitioned to rule on a noise complaint in 1722 was to ask the town clerk, firstly whether the Court had the power to take such a decision and, secondly, what records its predecessors had left as to what—if anything—they did when faced with noise-related complaints. Apparently no such records were found, and it is not known whether the aldermen did decide whether they were allowed to make a decision or not. In any case, they don't actually seem to have made one.

The complaint that so flummoxed the aldermen was about a coppersmith, and it is the noise of this trade more than any other that is singled out for comment in the early eighteenth century. This is partly because it really is a horribly noisy business: copper, unlike iron but like silver and gold, is beaten cold—yet it is the hardest metal so worked, and hence the loudest. And, as no forge was required, coppersmiths could work from home. Finally, there was a great increase in the trade at this time—and as usual the novelty of the sound made it particularly annoying.

In general, throughout modern history, the law has been uneasy about dealing with noise, and it has been suggested (particularly by Karin Bijsterveld in her 2008 book *Mechanical Sound*) that there are three main reasons for this. The first is the necessity until the mid-twentieth century of relying on subjective reports of the amount of noise involved. Secondly, there has long been a widespread acceptance that people have both some kind of a right to express themselves and at the same time some kind of a right to peace. The third reason is the realization that what bothers one person does not necessarily bother another.

When noise *was* the subject of litigation in the eighteenth century, it was always in some special circumstances—never simply about its impact. For instance, in 1724, Lady Arabella Howard took legal action against a group of churchwardens and others because of the 'very ill Consequence' of the local church bells in Hammersmith being rung at 5 a.m.[9] However, the justification for the action was not the noise itself, but the claim that Lady Howard had paid a previous churchwarden not to allow such early ringing.

If one were to judge the severity of noise problems simply in terms of official records of concern, they would not seem very important: there is no mention of noise in the rather dauntingly entitled *A Briefe Declaration for What manner of Speciall Nusance concerning private dwelling Houses, a man may have his remedy by Assize, or other Actions as the Case requires* (1639),[10] nor is noise much of a problem according to a book about *Public nusance, considered under the general Heads of Bad Pavements, Butchers infesting the Streets, the Inconvenience to the Public occasioned by the present Method of billeting the Foot Guards, and the Insolence of Household Servants, with some Hints towards a Remedy and Amendment*, the title of which, while not exactly snappy, usefully summarizes what its author, 'A Gentleman of the Temple', thought were the main causes of concern in 1754.[11]

It would be astounding if there were really no neighbourhood noise problems in this period, especially in older and cheaper housing stock, where conditions were cramped and walls were of an incredible thinness—often even party walls were simply wainscot (single-thickness wood panel) partitions. The growth of London and its noise was continuing apace, so much so that Alexander Pope, who had escaped to the peace of Twickenham, referred to the city as a New Babel. And it is quite clear from other evidence that noise was of concern, if not to the Gentleman of the Temple: there is no better example of the desperation to which some were driven by it than the fact that the Church of St Mary-le-Strand, completed in 1717, was constructed with no ground-floor windows at all, specifically in order to exclude noise.

Noise in the streets

In European cities, around the mid-eighteenth century, the rising tide of noise outside homes led in some cases to the movement of studies and offices into the central areas of buildings. It may seem surprising that such major structural changes could be made while such simple expedients as double glazing were not attempted—but there was still no science of building acoustics to call on for advice. Science was being applied to noise-*making*, however, and, thanks to Morland and Kircher, mobs now had 'Stenterophonic tubes' to make their demands more plainly heard.

By day, the streets of eighteenth-century European cities were very different from those of later eras: for one thing, there was a constant jostling for position by pedestrians and vehicles alike to gain the relative safety and cleanliness of the edges of the street and to avoid the central faeces-filled open drains. There were also far more types of street-user than there are today, many of whom, from beggars to entertainers, were not only themselves not in motion, but were actively trying to stop those who were.

Meanwhile, iron-shod horses pulling iron-tired vehicles over cobblestones meant that US cities at the time were already dominated by traffic noise in many areas. In Philadelphia, Benjamin Franklin described the 'thundering of coaches, chariots, chaises, wagons, drays, and the whole fraternity of noise'.[12] In Boston in 1747, a similar clamour led the Great and General Court of Massachusetts to ban all traffic from either side of the State House while it was in session.[13]

A German visiting London in 1770, Georg Lichtenberg, described the scene around Cheapside and Fleet Street in this way:

> crash! a porter runs you down, crying "By your leave", when you are lying on the ground. In the middle of the street roll chaises, carriages, and drays in an unending stream. Above this din and the hum and clatter of thousands of tongues and feet one hears the chimes from church towers, the bells of the postmen, the organs, fiddles, hurdy-gurdies, and tambourines of English mountebanks, and the cries of those who sell hot and cold viands in the open at the street corners. Then you will see a bonfire of shavings flaring up as high as the upper floors of the houses in a circle of merrily shouting beggar-boys, sailors, and rogues. Suddenly a man whose handkerchief has been

> stolen will cry: "Stop thief", and everyone will begin running and pushing and shoving—many of them not with any desire of catching the thief, but of prigging for themselves, perhaps, a watch or purse. Before you know where you are, a pretty, nicely dressed miss will take you by the hand...Then there is an accident forty paces from you...Suddenly you will, perhaps, hear a shout from a hundred throats, as if a fire had broken out, a house fallen down, or a patriot were looking out of the window...I have said nothing about the ballad singers who, forming circles at every corner, dam the stream of humanity which stops to listen and steal.[14]

By the mid-eighteenth century the noise of London—probably the world's noisiest city at the time—started to change in a fundamental sense. While previously annoyance and disturbance had been caused by a variety of individual noises, from street criers and musicians to pigs and dogs, now the sounds were all merging together—for the first time it becomes appropriate, at least at some times and places, to talk about noise levels rather than noises. From afar, the whole of London was said to make a humming sound, and, within, its inhabitants were often bathed in a sea of sound, with individual elements hard to isolate or identify. This change was ushered in partly by the beginnings of the Industrial Revolution, through its introduction of machines that acted as continuous noise sources. But the main reason was simply the very crowded state of London's streets in this period, which was in turn due to the enormous increase in the urban population. This growth was not caused by the birth rate, which did not even keep pace with the death rate engendered by high infant mortality and poor sanitary conditions. The key to the population increase was the very large amount of immigration from the countryside and overseas. Some estimates suggest that over 8,000 people net per year immigrated into London in this period, and that the population grew from about 200,000 in 1600 to 575,000 in 1700 and 900,000 in 1800 as a result.[15] This influx was in turn caused by the great prosperity of London, which was due largely to international trade, especially the imports of sugar and tobacco. Meanwhile, London was exporting more and more goods each year, and by now fully a quarter of the city's workforce earned their livings directly from the very noisy port.

Not only were the streets not built to cope with such large numbers of people, but also many of the smaller streets were effectively private spaces where few outsiders would dare to venture because of the high levels of crime within them. And so the pressure on the public thoroughfares was increased still further.

And London streets had many more uses than travel and passing trade. Not only was music-making as popular an outdoor activity as ever (see Fig. 14); there was also a rapid growth in the numbers of shops, coffee houses, and taverns, many of which offered a wide range of refreshments. Business and rendezvous of all kinds were conducted in these places and, since many of the refreshments were

FIGURE 14. Hogarth's *The Enraged Musician* (1741), perhaps produced as a result of the widespread comments that foreign visitors by now found London to be an intolerably noisy city.
Dover Publications.

alcoholic ones, public drunkenness also increased in this period, together with the noise it caused.

Meanwhile, in America, noise was acquiring a sinister reputation as a transmitter of seditious information. On 9 September 1739 there was a rebellion among slaves working near to Charleston, in South Carolina. Having captured a stock of weapons, they went on a march on which slave-owners were killed and plantations destroyed—and the marchers beat two drums as they went. According to contemporary reports, it was the sound of these drums that was the signal for other slaves to join in, and eventually there were about sixty men and women involved. In many states, the use of drums by slaves was subsequently banned as a result.

Art of noise

In the 1780s a new Mozart quartet began to become popular in polite society. It was called, most encouragingly for the purposes of this book, *The Dissonance*. In case you haven't heard it and are tempted to rush out and buy it, it is well worth it, but not because it sounds terribly dissonant—at least, not unless you compare it with Mozart's other quartets. Like many another artist who made significant changes that everyone else copied, from Giotto to Jane Austen, Mozart's use of dissonant chords is by modern standards really rather melodious and tuneful. It is in the nature of sound that its classification as noisy/dissonant or not depends on the hearers' expectations, and on what they are used to, which means that we really cannot hear the *Dissonance* to be as jarring as Mozart's' contemporary audiences did. In fact, dissonance in eighteenth-century chamber music is usually so extremely restrained that many people find it just a little bit dull.

If Mozart's dissonance quartet can no longer be heard as it was intended, the same cannot be said for one of the most famous works of his teacher, Josef Haydn. His Symphony 94 carefully lulls the listeners into a relaxed—even sleepy—state of listening, by playing a simple melody through once. The same melody is then repeated, even more quietly, but, instead of finishing on a quiet note, the orchestra makes a sudden loud fortissimo—to make 'all the women scream' according to Haydn (though it has often been said that he did it because of rumours about people falling asleep at his concerts).[16] It must have been highly successful at the time, and it still works very well today—recent releases of recordings of the symphony have even been criticized for recording the startled reaction of the audience as well as the music. And this is despite the game being given away by the symphony being universally referred to in English as *The Surprise* (the German version is *Symphony with the Drum Beat*, which is hardly less of a spoiler). This just goes to show that what surprised then surprises now—a sound that is loud, sudden, and unexpected.

One contemporary critic read rather more into it than Haydn intended: 'The Surprise', he wrote, 'might not be inaptly likened to the situation of a beautiful shepherdess who, lulled to slumber by the murmur of a distant waterfall, starts alarmed by the unexpected firing of a fowling-piece.'[17] Or, on the other hand, perhaps it might be.

In themselves, noises have long had supernatural connotations, from the Irish banshee that is simply a scream or wail that warns of approaching death, to the moans of ghosts and the rattling of chains. Séances, too, were replete with strange noises—the tapping of tables, the whispers of the departed, the strains of angelic music—even the rattle of tambourines. In 1762 one of the best known of all ghost stories was reported, the Cock Lane ghost, for which the evidence was entirely acoustic. Knockings and bangings were heard, supposedly caused by a spirit in possession of a young girl who lived there. It was said that she was 'constantly attended by various noises'. It was eventually decided that the ghost was a hoax.[18]

Though they were attracting new interest in the late eighteenth century, the supernatural aspects of sound were not new: as long ago as the first century, Pliny the Younger told a supposedly true story of a haunted house. The basic details have changed little in the many centuries since, with the arrival of the ghost prefaced by the approaching rattle of chains.

In tales of fictional terror, noise is also frequently used to set the scene for ghostly occurrences, as in the howling gale and wintry blasts so often encountered in gothic fictions of the eighteenth century, pre-eminent among which is Anne Radcliffe's *The Mysteries of Udolpho* (1794), in which her heroine must cope with a range of supposedly supernatural menaces. She also has to put up with some terrible weather: 'The night was stormy; the battlements of the castle appeared to rock in the wind, and, at intervals, long groans seemed to pass on the air, such as those, which often deceive the melancholy mind, in tempests, and amidst scenes of desolation.'[19]

In general fiction too, now a well-established and very popular art form, a new source of information about noise becomes available to the historian: the novel. Tobias Smollett, in his 1771 comic best-seller *The Expedition of Humphry Clinker*, gives a vivid account of the noise of the city of London—by night and day:

> I go to bed after midnight, jaded and restless from the dissipations of the day—I start every hour from my sleep, at the horrid noise of the watchmen bawling the hour through every street, and thundering at every door; a set of useless fellows, who serve no other purpose but that of disturbing the repose of the inhabitants; and by five o'clock I start out of bed, in consequence of the still more dreadful alarm made by the country carts, and noisy rustics bellowing green pease under my window.[20]

Later in the book, the irritations and tribulations of an invalid in search of peace and quiet called Bramble are recounted at length. Like Morose before him, Bramble is mainly a figure of fun, but also, in part, of sympathy—the sympathetic aspect being enhanced by his illness. By now, the idea that those who are unwell are especially sensitive to noise was well established. Such views sometimes crop up in odd places, like Robert Clavering's 1779 essay 'On the Construction and Building of Chimneys', in which he says: 'in high winds nothing can be more disagreeable to a delicate and sickly person than the horrible noise the wind makes in whistling round the chimney pots.'[21]

7

MACHINERIES OF NOISE

Of all the many inventions of the eighteenth century, perhaps none did more to add to the level of ambient noise in European cities than the steam engine. Though some steam engines had been in occasional use since Thomas Savery's 1698 patent for 'A new invention for raiseing [*sic*] of water and occasioning motion to all sorts of mill work by the impellent force of fire', it was the technological advances of James Watt in the 1770s that introduced a vast improvement in efficiency over the Newcomen devices that were available at the time, and led to their great proliferation. Watt's engines were significantly better largely because he designed them—as far as he could—on scientific principles, rather than proceeding by trial and error and then stopping when a half-decent engine had been constructed and building lots more, as Newcomen did. As a result, steam engines were transformed from clumsy contraptions that were marginally useful, at the expense of an enormous output of wasted energy, to relatively efficient, carefully designed devices that revolutionized industry, and soon transformed the soundscapes of cities all over the world too.

It was not simply the hiss of the escaping steam that made steam engines noisy—a badly regulated or poorly built engine filled the air with a whole range of screeches, clanks, and groans as well. And these weren't all accidental: the first applications of Watt's improved steam engines were in Cornish mines, where they were employed to pump out water. But Watt and his newfangled invention were initially viewed with great suspicion by the miners.

What really helped to put Watt on the local map, made his engine popular, and even allowed him to gain a measure of grudging respect among his first customers was not so much the efficiency and effectiveness of the new engines as the awful noise they made:

> The velocity, violence, magnitude, and horrible noise of the engine [wrote Watt] give universal satisfaction to all beholders, believers or not. I have once or twice trimmed the engine to end its stroke gently, and to make less noise; but Mr Wilson [site manager] cannot sleep unless it seems quite furious, so I have left it to the enginemen; and, by the by, the noise seems to convey great ideas of its power to the ignorant, who seem to be no more taken with modest merit in an engine than in a man.[1]

Sounds of war

The 1780s, much like many other decades of the eighteenth century, were a time of wars, many of them naval, characterized by much heroism, a lot of bloodshed, and some very loud noises—so loud in fact that they caused hearing problems of such seriousness that the medical profession could hardy help noticing them. But noticing was all they did—observations made at the time were simply logged in diaries and records, and not even thought worth the bother of publishing, let alone doing anything about. In fact, the records that do exist saw the light of day only in the 1820s, in the form of posthumous publications of the work of a Dr Parry[2]—though even then, judging by the amount of concern and action the publications generated—which was none at all—the publisher might as well not have bothered. It seems safe to assume that people just were not that interested in a few deaf sailors, especially when there were many more dead ones. It is also fair to say that the sort of diseases, conditions, and privations that the average sailor had to endure probably made deafness a minor concern. It was only when famous people or events were involved that any note was taken of such things. So there were references to gun crews in the Battle of Trafalgar and Copenhagen, and to the hearing of an admiral: Lord Rodney, who became almost completely deaf after listening to eighty broadsides being discharged from his ship, the HMS *Formidable*. One can only wonder what the effects on the gunners were. Those in the Trafalgar cases were made completely, instantly, and permanently deaf.

Artillerymen and others involved in land battles had their hearing damaged by the roar of canon and muskets, just as their marine equivalents did. The Duke of Wellington's hearing loss in his left ear was apparently initiated by artillery fire at the Battle of Waterloo, but at least part of his later problems with it was due to his doctors, who administered a whole range of treatments that were at best useless and at worse dangerous. The effects even made their way to the pages of the *London Times*, which, in its 25 September 1822 issue, reported: 'A violent remedy, applied to or poured into the ear, to cure a temporary deafness, or perhaps the gradual approach of that infirmity, is said to have wholly affected his head, and to have produced excessive agony and fever, so that at one time his medical attendants are rumoured to have apprehended disastrous consequences...'. It was as a result of such treatments that Wellington became completely deaf in that ear. Ludwig van Beethoven—whose deafness was probably caused by otiosclerosis—fared little better. One of his doctors prescribed a 'kind of herb for my ear. Since then I can say I am feeling stronger and better, except that my ears sing and buzz constantly, day and night.'[3]

Beethoven, driven crazy by his deafness and other ailments, did in fact develop his own, partially effective solution—by biting on a drumstick, he was able to hear a little when he pressed the other end to a hard surface. Later, Thomas Edison, also plagued by deafness (caused by scarlet fever), came up with a similar solution, and mementoes of this survive in the form of tooth marks on his personal phonograph. Today a more high-tech version of this solution is used by divers: while they usually speak to each other through microphones, the replies are sent from a helmet-mounted receiver to a rod that the diver bites on, from which the sounds travel to the inner ears through the bones.

The use of noises to assist in medical diagnosis also took a significant step forward at this time. Though doctors had been listening to the noises of their patients' digestive systems, lungs, and hearts—for thousands of years, contemporary male medics were reluctant to press their ears to young ladies. As impressively named French physician René Théophile Hyacinthe Laennec put it, the procedure was 'rendered inadmissible by the age and sex of the patient'.[4] Instead, Laennec rolled a piece of paper into a tube, stuck one end on the lady in question, the other in his ear, and the stethoscope was born.

Sound in theory

The acoustically related work of scientists in the nineteenth century began in earnest in the 1820s, and much of the work done in this decade was to have significant impacts in the longer term.

One example is the measurement of the speed of sound underwater. Today, this is vital for all sorts of applications, from global warming research and weather forecasting to fish-stock assessment and seabed mapping, but the impetus for the experiment that first determined it was a prize offered by the French Academy of Sciences. The prize was not for acoustics at all, but for the measurement of the compressibility of a range of liquids. Swiss physicist Daniel Colladon decided to compete by measuring how compressible water is, and he was surprised by the smallness of the answer: a pressure increase of 1 atmosphere reduces the volume of water by only 0.0053 per cent. Since he knew that the equation for the speed of sound predicts that, the less compressible a fluid is, the faster sound passes through it, Colladon realized that the speed of sound in water must be very high indeed.[5] But no one knew what it was. So, having plenty of time on his hands and an interest in messing about in boats, Colladon decided to find out.

For Colladon's first experiment, he and his friend Monsieur de Candolle (along with Monsieur de Candolle's gardener, in case there was any actual work to be done) took a pair of boats out on Lake Geneva. They rowed about a kilometre apart, and the gardener suspended a bell just below the surface of the water. He then struck it with a clapper while M. Candolle waved to M. Colladon and started a stopwatch. As soon as he saw the signal, Colladon plunged his head into the water, and then lifted his arm when he heard the sound of the bell, at which point Candolle stopped the watch. Since it was a breezy day, the combination of a rocking boat and the need to wave his arm in the air while keeping his head under water led him to wonder whether there might be a better approach.

So Colladon built the world's first hydrophone, which turned out to be rather easy. It consisted of a watering can (presumably belonging to the gardener) with weights in it. Sure enough, sounds that travelled through the water could be clearly heard via the can's spout, and Colladon's head remained completely dry this time. So, he had a purpose-built version made—a tube with a closed spoon-shaped vessel at the end—and set off eagerly to use it (see Fig. 15).

Colladon was keen to increase the distance between the bell and the hydrophone as much as possible, since he would get a more accurate value for the speed in that way: his stopwatch measured to the quarter-second, which would lead to an uncertainty of around 40 per cent for a kilometre but only about 3 per cent if he used the entire length of Lake Geneva, just over 14 kilometres. Colladon called in his father and sent him to the village of Roll, with strict instructions to find a boat with a 12-metre mast with a lamp on top, equipped with a cover that could be removed quickly.

Colladon's account is a meandering one, suggestive of a period when the pace and pursuit of scientific research were more relaxed, with many picturesque details included. Even when we are at last on

Bateau expéditeur du son.
Figure 1

Bateau récepteur du son.
Figure 2

FIGURE 15. Colladon's experiment, from his 1827 account in *Annales de Chimie et de Physique.*

J. D. Colladon, *Souvenirs et mémoires* (Geneva: Albert-Schuchardt, 1893).

our way with him to Lake Geneva, armed with some fireworks that he planned to set off so that his father would know when to bang his bell, we are delayed, just as Colladon was, by an officious customs agent who was alarmed by the fireworks. Finally, however, he allowed Colladon to set up his equipment and send off a firework. As instructed, his father rang the bell on the other side of the lake and uncovered the light on the top of the mast—but Colladon couldn't see it. Colladon senior had only managed to find a boat with an 8-metre mast and had consequently been foiled by the curvature of the Earth.

So, the experiment had to be tried a third time. For this attempt, Colladon's father was given the even more challenging task of simultaneously striking a bell and setting light to a pile of gunpowder large enough to make a flash at least 12 metres in the air, while at the same time not blowing himself up.

Surprisingly enough, Colladon senior not only concurred with this plan but carried it out too, and the experiment was a great success. Colladon obtained a value of 1,435 metres per second, which is in amazingly good accord with modern values (1,438.8 metres for second, for the appropriate temperature).

While Colladon was experimenting with his primitive hydrophone, across the Channel Charles Wheatstone, co-inventor of the electric telegraph, was developing its airborne equivalent, the microphone. Its purpose was, to allow people to hear the inaudible (the source of the name is the similarity of this concept to that of the microscope, a device allowing the user to see the invisible).

Despite the fact that Wheatstone's telegraph was one of the first practical devices to be based on the principles of electricity, his microphone was not of the plug-in variety, but purely acoustical in nature. It was simply a pair of metal plates, one for each ear, joined together at one corner. A springy metal rod was attached to the opposite corners, to pull the plates together. One put one's head between the plates, which the rod pulled together like a fortunately not very strong vice (I say 'one' here, but there's no record of anyone actually doing this but Wheatstone himself). To avoid the microphone slipping off, a dressmaker's ribbon was also employed to hold the apparatus in place. Unfortunately, no images of Wheatstone wearing his invention have survived.

Clearly thinking that the uses of such an apparatus might not be immediately evident to his readers, Wheatstone suggested what might be done with it. Among other things, it could be used as a dissonance detector, simply by singing while striking a tuning fork and touching the plates with it (so that 'the most uninitiated ear will be able to perceive the consonance or dissonance of the sounds'). Another of Wheatstone's suggestions was to press one's head to a vessel of boiling water while wearing the microphone, so that the sounds made by boiling water could be listened to—though presumably not for very long.

The microphone sank without trace—all but its name, which was purloined by Arthur Hughes when he invented the electrical version half a century later.

Meanwhile, also in England, more fun with science was being had in Sir Henry Bunbury's library, in Suffolk. There William Hyde Wollaston, famous discoverer of two new elements and inventor of the *Camera Lucida,* had taken to concealing himself behind the book stacks. He was equipped with a series of tiny pipes, carefully constructed to be as high-pitched as he could make them. He waited for unwary book-lovers to settle themselves in the library and then blew sharply on the pipes, while watching carefully to see if the readers jumped. The result was his 1820 paper 'On Sounds Inaudible by Certain Ears', which laid the foundations of the experimental study of the range of pitches that humans could hear and hence of the science of ultrasonics.[6]

Noise in practice

In searching for the publication of anything that was actually of any immediate use for the understanding or control of noise, only a single contribution emerges for the first half of the nineteenth century: a slim volume published in 1826 by Henry Matthews. Its title reflects the main areas of acoustic concern at the time: *Observations on Sound; Shewing the Causes of its Indistinctness in Churches, Chapels, Halls of Justice, &c with a System for their Construction.* Explaining that 'improvement in the science of sound is of consequence as respects a still more sacred subject than even justice itself—Divine Knowledge', Matthews goes on:

> It is certain that in churches constructed as they now are, all sound that is confined is only rendered mischievous, instead of useful; or at least nineteen parts out of twenty. For not more than one part out of the twenty proceeds in a direct line to the audience. In the open air, the nineteen parts are lost, but in our present churches &c, they are the enemies of the one good part.[7]

In many ways, the analysis that Matthews makes of the problem is spot on, and the pamphlet was the first to account clearly for the causes and effects of reverberation, explaining very graphically:

> That part of the sound which proceeded in a direct line would arrive first; and that reflection from an oblique wall, having to perform a longer journey, would arrive later . . . Echo does not politely wait until the speaker is done; but the moment he begins and before he has finished a word, she mocks him as with ten thousand tongues.[8]

Matthews also explained correctly just how and why sermons go over the heads of congregations: the warm air rising from the people changes the density of the air above them and deflects the sound from the pulpit upwards.

Anyone encouraged by the last part of the book's title to hope for effective solutions to the problems described so accurately in it was to be disappointed. Various modifications to the shapes of church interiors are suggested, designed to reflect those mischievous nineteen out of twenty parts away from the congregation until only the quiet but pure twentieth part remains—but none of them is likely to have made any significant difference. As we now know, the problem with reverberation is that reflecting sound in different directions does little good; what is needed is to absorb it.

Regarding the over-their-heads problem, Matthews suggested using air draughts to blow the sound back to the congregation. It is none too clear how these could have been arranged, nor how the congregation would feel about being required to worship in a constant draught, but they wouldn't have worked either.

The absence of publications about noise control in the period should not be taken to indicate a lack of concern about the problem. One area that was causing great perturbation to some in the 1820s was the rise of the railways. The world's first commercial railway line, between Stockton and Darlington, was opened in 1825, and in 1830 the Manchester to Liverpool line followed. Other countries were not far behind: in particular, there was rapid development in the USA from 1828, which continued steadily for the next five decades.

Early UK railways were pushed forward by the tenacity and pragmatism of George Stephenson in the face of enormous opposition from landowners, many of whom referred to the anticipated noise of the railways.

Generally speaking, it is thought today that a lot of the opposition to railways (which included mugging railwaymen and blocking routes) arose simply because landowners were not happy to have to sell any part of their land, but it is possible too that Stephenson's vision had exceeded his skill: his first locomotive, built in 1814 and called the *Blücher*, was just about as quiet as it was attractive (see Fig. 16). According to contemporary account:

> It was exceedingly uncouth and cumbrous in appearance, as well as unsteady in its motion. Being without springs, it jolted and jerked along in a fashion that was not calculated to give the beholder a favourable impression regarding its powers. The steam after passing through the cylinders, escaped with a horrible noise, which caused the colliery owners to be threatened with law proceedings for the terror produced by the awful machine upon cattle and horses.[9]

However, Stephenson was a determined man and persevered in his attempts to build a better locomotive. His next effort was rather better engineered—but its name, *Puffing Billy*, was not a very accurate one. It didn't so much puff as shriek. As another contemporary

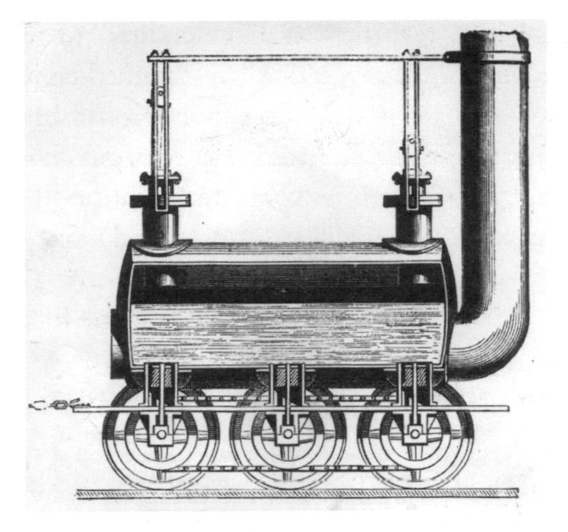

FIGURE 16. The 'uncouth and cumbrous' *Blücher*, 1814.
Hulton Archive/Getty Images.

account puts it, 'the jets of steam from the piston, blowing off into the air at high pressure while the engine was in motion, caused considerable annoyance to horses'.[10]

The first solution to this problem was not a technological one—the driver and guard were instructed to keep their eyes peeled for horses and to stop the locomotive as soon as they saw one. Not very surprisingly, 'much interruption was thus caused to the working of the railway, and it excited considerable dissatisfaction among the workmen'.[11]

So that idea was given up and a more scientific approach was adopted: the exhaust steam, instead of being vented noisily into the air to annoy the local livestock, was instead ducted to an internal reservoir behind the chimney, and it gradually—and rather less noisily—escaped from there. As it turned out, this design also greatly improved the efficiency of the engine—a rare example for the time of the remedy for an acoustic problem having a non-acoustic pay-off.

Stephenson's triumph, the *Rocket*, which was built in 1829, won the Rainhill trials, the competition to decide which locomotive would be allowed to run on the Manchester–Liverpool line. In reporting on the trials, the *Mechanics Magazine* referred to 'the absolute absence of all smell, smoke, noise, vibration, or unpleasant feeling of any kind, the elegance of the machinery, in short the *tout ensemble* proclaim the perfection of the [locomotive] principle'.[12]

A continuous roar

By the early part of the nineteenth century in London and other industrial cities, the change in the character of noise that had begun a generation earlier was complete: as the extent and pace of the Industrial Revolution increased, noise became anonymous, as individual sounds became first a cacophony and then a level—a 'hum' referred to frequently from then on. This change had several effects, beyond the obvious one that it became much harder to track down and deal with individual noise sources.

For one thing, people get used to continuous noise much more easily than to isolated sounds: there is no startle effect, and they become habituated to it. However, continuous background noise—even if it is no longer consciously heard—has a wide range of negative effects on people, from stress, tiredness, and increased irritability to reduced accuracy in task performance. For those working very close to the noise sources, continuous sounds are relatively less damaging than short-duration ones, as the ear's tensor tympani and stapedius muscles have time to respond to the level by tensing the eardrum and partially immobilizing the ossicles respectively.

Another effect was that, in noisy workplaces, work songs were no longer possible, the extreme noise levels limiting communication to what was absolutely essential because of the need to bellow at the top of one's lungs, and so in consequence each worker became increasingly isolated (though workers were often banned from speaking to each other in any case).

A larger fraction of the population was now being exposed to high levels of noise, partly because of the enormous growth in industrial production, which affected many regions. The production of coal, for example, a mere 6.2 million tonnes in 1770, had risen to 11 million in 1800. Compared to that level, it doubled by 1825 and tripled by 1840: by 1850, it was more than four times as large, at 49.4 million tonnes.[13] Pig iron output and cotton imports (used for textile production) grew at similar rates.

In view of these changes, it is not surprising that the medical community at large began at last to recognize that noise was a health problem: the first authoritative reference appeared in 1831, when Dr John Fosbroke, writing in the *Lancet*, stated that 'blacksmiths' deafness is a consequence of employment', adding that the condition 'creeps upon them gradually'. Though hardly describing a new discovery, this paper marked the beginning of a fairly widespread effort by physicians to understand the details of noise-related health problems.[14]

For the time being, the newfound interest in noise-related health problems—such as it was—remained within the medical community. It is perhaps not surprising that factory-owners ignored the problem, but one might have hoped for better from those assigned to assess factory working conditions. Sadler's Factory Investigating Committee was the first to report (1832) on the conditions that factory workers endured in London factories. Though 700 pages long, the report contains scarcely a mention of noise throughout. In the brutal context of routine overwork, widespread abuse, and severe punishment, noise must have seemed a very minor problem, but, on the other hand, since it is one that the visiting committee members must have been exposed to themselves, it is surprising that it should not have been referred to. Some interviewees even mention the 'silence' of the mills, referring to the rule that the workers must not speak to each other. It is noticeable that the upcoming noise campaigns of the nineteenth century hardly mentioned factory noise either—partly, of course, through lack of familiarity with it, though Charles Dickens, one of its most enthusiastic supporters, was well aware of the issue, having had firsthand experience: 'I am taken to the Cutting Room . . . and subjected to the action of large rollers filled with transverse knives, revolving by steam power upon iron beds . . . Such a drumming and rattling, such a battering and clattering, such a delight in cutting and slashing, not even the Australian part in me has witnessed before.'[15]

It may be that noise was so intrinsically associated with industrialization that it seemed to be an irremovable part of it, and that to criticize noise would be like criticizing steam power itself. Meanwhile, at least in Dickens's view, noises that connoted religion were becoming increasingly unwelcome, as those of factory machinery overlaid them and as the focus of life gradually shifted towards that of industrially derived commerce:

> It was very strange to walk through the streets on a Sunday morning and note how few of them the barbarous jangling of bells that was driving the sick and nervous mad, called away from their own quarters, from their own close rooms, from the corners of their own streets, where they lounged listlessly, gazing at all the church and chapel going, as at a thing with which they had no manner of concern.[16]

Certainly, complaints about industry as such were few and far between, perhaps showing that by now industrial machinery had fully supplanted religious sound sources as 'sanctified' or 'sacred' ones. And in fact, as the nineteenth century progressed, complaints against church noises do begin to appear—such as an 1896 petition in Aurillac, France, against the sound of a particular church bell, and an English lawsuit of 1851, *Crump* v. *Lambert*, which resulted in a restraining order to stop one small church (adjoining Crump's home) ringing its bells. And J. H. Girdner, a New York physician eager to provoke an anti-noise campaign, wrote in 1898 that, 'in these days of innumerable clocks and watches, the ringing of church bells in large cities is simply barbarous'.[17] One might speculate too that the rise in noise complaints of all kinds since the middle of the twentieth century has been caused at least in part by a new shift in power, from industrial organizations and governments to individuals and to pressure groups. This would also help to explain the reluctance of authorities to deal with noise made by individuals, even though they act speedily against noisy factories. This type of approach to the history of noise is that of Murray Schafer in his 1977 book *The Soundscape,* while the concept of noise as potentially sacred originates with anthropologist Claude Lévi-Strauss.[18]

Noises of road and rail

For all the neglect of noise problems in most cases, one area in which it became a major factor was in the competition between transport technologies for road use. Unlike railway locomotives, these vehicles had to share their routes with horses, so it was essential that they did not disturb them—and, ideally, not upset their own passengers too much either. For engineer and inventor Walter Hancock, quietness was the guiding principle in developing his steam car (see Fig. 17). Hancock wrote in his *Narrative of Twelve Years' Experiments* of 1838 that his machines caused 'no annoyance by either noise or smoke to either bipeds or quadrupeds'.[19] Indeed, they were supposedly so quiet that it was said that horses came and looked in the cab to see how they worked.

Though in their heyday of the mid-1830s Hancock had three steam carriages running in London, ultimately they, like all other attempts to construct steam-powered road vehicles, were stymied by the ruinous state of the roads at that time. The fact that the vehicles had seven(!) radiators, which exploded from time to time, may also have been a factor in their abandonment. The future, it was soon clear, ran on rails.

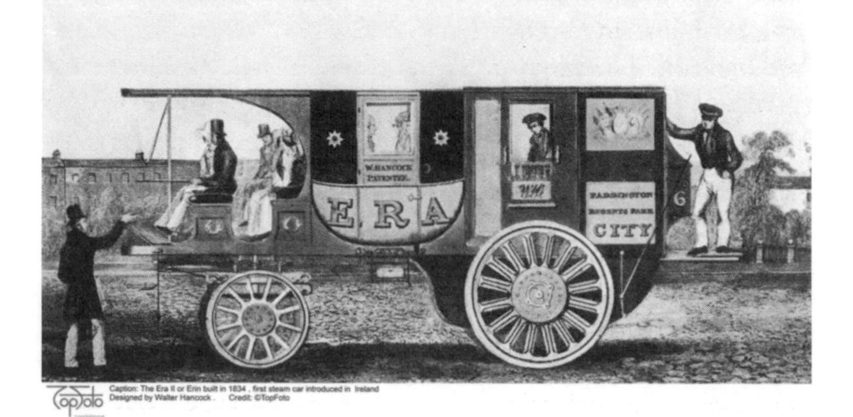

FIGURE 17. Walter Hancock's steam carriage *Automaton* (1827), which, for a few months, provided a regular service between London Wall and Paddington. TopFoto.

More than any other period in history, in the UK the 1840s was the time of the railway. A frenzy of track-laying transformed the country in the decade from 1840, when a few scattered lines linked major towns, to 1849, when a vast rail network had spread weblike over much of the country and linked many towns and villages into one interconnected system. Money from investors poured in, local rural economies were transformed, and people began to travel as never before. The change in the soundscape was enormous, especially in country areas. Quiet villages were suddenly host to roaring assemblages of steam-powered iron thundering down metal tracks and splitting the air with the shriek of their whistles.

No one had heard the thunder of wheels on iron rails before, other than the squeak and rumble of a few horse-drawn wagons in close proximity to mines, but, despite the newness of the noise and the rapidity of the change, the number of complaints, official or informal, about the railways was very small. And small it was to remain: the report of a major survey carried out in 1963, just before the Beeching Report greatly reduced the number of stations and the length of line in the UK, concluded: 'There is no evidence of widespread public annoyance from railway noise.'[20]

Why is this? Well, it does seem to be the case that, for some reason, people just don't mind too much about the noise of trains—they didn't in the 1840s and they don't now. So well established is this phenomenon that it has a name—the 'Railway Bonus'. Planners often give a 5 dB or even a 10 dB reduction to their estimates of the noise caused by railways to indicate that people are about as bothered as they would be by road traffic noise 5 or 10 dB greater.

Several explanations for the bonus have been suggested. For one thing, trains are predictable. Your local line might wake you up at 11 p.m., but at least you know it *will* be at 11. And railways are well behaved in space as well as time, sticking obediently to their tracks and never popping up in unexpected places, as—say—helicopters do. Perceived safety, too, is an issue: when a plane flies overhead, you're never absolutely sure that it isn't suddenly going to fall on you, but trains rarely kill people who aren't travelling on them. And, for some people, the sound of a train chinking over its ill-fitting rails in the distance of a summer night has a nice homely, nostalgic feel (hence the popularity of *The Railway Children* and *Thomas the Tank Engine*).

People of the 1840s would have felt no such nostalgic affection, of course. Though in 1844 Nathanial Hawthorne complained that the noise of the train spelled the end of rural peace, for many the world over an end to rural peace was just what was wanted. At last they could travel easily to towns, for business or for pleasure, and the journey itself, at breathtaking speeds in excess—occasionally—of 30 mph was itself an exciting adventure. Perhaps to them the roar of a train departing was not dissimilar to the blast-off sound of a 1960s Saturn V or a 1980s shuttle, making their way to unfamiliar worlds.

In the UK, one consequence of the rapid rise in transportation, the growth of cities, the mass movement of people, and in particular the need to find accommodation for the vast influx of mostly

Irish labourers in the 1830s and 1840s was the building of extremely poor-quality housing. In fact, the phrase 'Jerry-built' (perhaps derived from a sailors' word, 'jury', meaning makeshift or slapdash) is said by some sources to have been first used to describe them. Edwin Chadwick, in his 1842 *Report on the Sanitary Conditions of the Labouring Population of Great Britain,* says that 'the walls are only half-brick thick, or what the bricklayers call "brick noggin", and the whole of the materials are slight and unfit for the purpose. I have been told of a man who had built a row of these houses; and on visiting them one morning after a storm, found the whole of them levelled with the ground.'[21] Unfortunately, many were built sturdily enough to survive well into the 1930s (and, in a few cases, the 1960s), causing a continuing annoyance to those who dwelt in them.

Meanwhile in Belgium, perhaps because he noticed the distinctively changing note of the whistle of a passing train, Christiaan Doppler was completing his mathematical description of the effect that now bears his name, in which the waves sent by a moving object are measured as shorter or longer depending on whether the object is moving towards or away from the observer. Doppler was an astronomer and he formulated his law for light waves,[22] but, whether or not it was suggested by railways, it was certainly demonstrated by them, in an experiment carried out three years after Doppler had published his discovery. C. B. J. Buys Ballott, a Dutch meteorologist, talked the Rhine Railroad Company into allowing him to carry out an experiment that must have done its bit to convince people that scientists were all a bit odd. First, a band of trumpeters climbed aboard an open carriage, and played their instruments as continuously as they could while the carriage proceeded slowly along the railway between Utrecht and Maarsen. At intervals along the track, musicians were carefully arranged, and, as the carriage trundled past each one, he estimated the change in pitch of the trumpet note. Then the locomotive was sent back at a higher speed and the experiment repeated—and so on.

After the locomotive had showed off its highest possible speed, the musicians swapped places with the trumpeters, and it was their turn to be wobbled along the tracks, again estimating the change of pitch as they passed each trumpeter.

Finally, all the observations were collated and compared with the speed of the train. Despite the dependence on the judgement of the musicians, the tests were fully in accordance with Doppler's predictions.

In retreat

Just as in science, where developments in acoustics frequently spring from work intended for something quite different, so in law, regulations set up with some other aim are often the ones that prove most useful for the more specific concerns of noise. The very first national law that could—just about—be used to control noise was enacted in 1846, in England. The Public Health Act gave local health boards the power to supervise unhealthy or annoying factories, and could force a factory-builder to seek permission before proceeding with construction. In its original form, the Act was quite lacklustre, since no right was conferred on members of the public: only complaints from recognized officials were considered by it. Furthermore, complaints could be upheld only if the owner could be shown not to have adopted the most practicable way of preventing a nuisance. It was this Act that ultimately led to the raft of laws and regulation that protect people from noise today, but in its original form it was quite inadequate properly to tackle the problem of noise, which was steadily increasing as London's population grew. The 1840s was the decade of one of its greatest increases, with, according to some figures, a 26 per cent rise between 1841 and 1851 (from 1,870,272 to 2,362,236).[23]

One of the key features that distinguishes individual annoying noises from an annoying level of noise is that no one person is likely to be responsible for the latter—and, hence, the possibilities of legal action on an individual basis are very limited. This was just the problem that faced many at the time. As Thomas Carlyle said in 1824:

> You are packed into paltry shells of brick-houses; every door that slams to in the street is audible in your most secret chamber... and when you issue from your door, you are assailed by vast shoals of quacks, and showmen, and street sweepers, and pick-pockets, and mendicants of every degree and shape, all plying in noise or silent craft their several vocations.[24]

Carlyle was one of several writers commenting on noise in the mid-nineteenth century. Unsurprisingly, they focused on those noises and noise-makers that affected them directly, and they decided that 'brain workers' like themselves were more susceptible to the effects of noise than manual workers. The effects of this narrow view are both positive and negative—on the one hand, it led to the gathering-together of the like-minded into groups that were quite effective in campaigning for the control of noise sources; it is the writers of the nineteenth century who are the first to generate widespread criticism and action against noise. On the other hand, the very similarity of those campaigners made them deaf to the unwritten complaints of others who have their own least favourite noises to worry about, which meant that the campaigns were not broad enough in their application to generate widespread support, or to have any effect at all on those types of noise that did not affect intellectuals—in particular factory noise. They also had little impact on the general level of noise. In the case of Dickens, this parochialism is surprising, since he was well attuned to other public ills, whether they affected him directly or not. Perhaps part of the explanation is that noise was not to be regarded as a pollutant until over a century later and so people considered it as a strictly local and interpersonal problem.

Carlyle did more than simply write about noise: his response to the 'vast shoals' of noise-makers was to change the paltry shell of his own house by building a soundproofed study on top of it. While it was being built, he escaped to the peace and quiet of the country, leaving his wife to superintend the rebuilding work (which she actually quite enjoyed, commenting that the noise seemed much less bothersome when her husband was off the premises). When it was completed, in 1843, all was well to begin with—as he wrote to his mother, Margaret, in November of that year: 'My little upper room far out of the noise of pianos was finished near a fortnight ago... here I sit, lifted above the noise of the world, peremptory to let no mortal enter upon my privacy...'.[25]

But Carlyle's triumph was not to last—he soon found that he simply became more aware of the sounds made within his house, which was most irritating—as was the fact that his wife ran off with his architect.

The cause of the former of these annoyances would have been clear to Carlyle if he had heard of the work of Ernst Heinrich Weber, who in 1846 announced a key breakthrough in the understanding of the reaction of living things to stimuli of all kinds. Weber discovered that, 'when stimuli multiply, sensations add' or, more accurately, that the change in perception caused by a change in stimulus is proportional, not to the actual increase in the stimulus, but to the logarithm of that increase. So it might be that, as a stimulus becomes ten times stronger, the perceived increase only doubles—and only doubles again if the original stimulus becomes 100 times stronger. In 1860, Gustav Fechner's *Elements of Psychophysics* was published, elaborating on Weber's discovery and bringing it to a wider audience.[26] Weber's law (as Fechner called it), is very convenient for living things, as it means that they can accommodate perceptions of a vast range of levels of stimuli without being overwhelmed by them. And it has a number of important consequences for noise. For one thing, it makes the description of the strength of noise in terms of the obvious units—newtons per square metre for pressure, watts for power, or watts per square metre for intensity—both clumsy and not very useful, since doubling the values of such units does not represent a doubling in the perceived sound. It also makes the task of reducing the impact of environmental noise very challenging. If water is flowing into your houseboat through two identical openings, closing one will halve the inflow and give you twice as long before you sink. But close one of two open windows to exclude the noise of a party on your street, and you will hardly notice the difference—the sound power in your room has halved (assuming that all the sound was coming in through the windows), but you will only hear a reduction of about 3 dB—barely noticeable. A similar issue occurs if one tries to organize a minute's silence in a cheering football crowd. If 9,999 out of the 10,000 people fall silent, the one remaining person's voice will be highly noticeable, even though the total sound power will indeed have fallen by about 99.99 per cent.[27]

It is not just sound pressure levels to which the human ear and brain respond logarithmically—the same is true of frequencies, with the result that piano keys are spaced logarithmically, not linearly.

Hearing and deafening

While Carlyle was retreating from noise in England, and Weber was finding out exactly why such retreats are so tricky to pull off, in America the noise of the Civil War was attracting attention. The Battle of Gettysburg (1863) is recorded as particularly noisy, though perhaps only because it was the first to be described at length by those involved. Union General John Gibbon said of it: 'To say that it was like a Summer storm with the crash of thunder, the glare of lighting, the shrieking of the wind, and the clatter of hailstones would be weak.'[28] One-third of soldiers suffered hearing loss as a result of the war—which, for the first time in history, was classified as a service-related condition.

The whole course and outcome of this particular battle were defined by noise effects: lacking any form of electrical communication, commanders used the sounds of over-the-horizon battles to monitor their progress and deployed their troops according to the relative loudness of the battle noises from different directions. That this is a misleading approach is clear from the fact that Union General Meade could not hear the Gettysburg battle from his location at Taneytown, 19 kilometres distant—yet it was heard clearly 240 kilometres away in Pittsburgh. Such effects were key the following day—Confederate General Ewell did not move his troops because he didn't hear the artillery of his fellow general, Longstreet. Lacking these troops, Longstreet was defeated by Meade.[29] The cause is the same as the one that caused Matthews's church-goers not to hear their vicar: heated air refracted the sound waves over Ewell's head.

A great stumbling block in the assessment of nineteenth-century noise is that, though the increasing pace of the Industrial Revolution and the growth of cities, together with the written records of the time, make it clear that noise increased substantially, there is no way to determine this increase quantitatively. However, it was soon to be possible to do this, and to distinguish vociferous complaint from actual change in noise level, as more scientists turned their attention to sound. By far the most important in terms both of his long-term contributions and of his influence at the time was Hermann Helmholtz, a fine scientist and something of a romantic. Helmholtz's ground-breaking 1856 book (translated into English in 1885), *On the Sensations of Tone*, laid the foundations for several areas of modern acoustics. Taking advantage of the fact that acoustics at the time was a very ill-defined topic, Helmholtz ranges over many subjects, aesthetic as well as scientific. In particular, his explanation of the mechanism of the ear as a set of nerves, each of which resonates in response to a particular tone (Helmholtz referred to it as a 'nervous piano'), while not correct, was on the right lines, and was highly influential. It is Helmholtz too who offered a basically sound explanation of the cause of the unpleasantness of dissonances. He rightly suggested that the ear is an analyser: presented with a complex sound, it works out how that sound may be constructed by adding together individual, simpler sounds.[30]

Helmholtz suggested that, if we hear a sound that the ear analyses into component sounds (which he called partials) that are so close as to produce beats, then that sound will be a dissonant one—especially if the beats occur at a rate of 30–40 per second.

Helmholtz's figures were right, but he could not give a complete analysis because he was unaware of the detailed workings of the ear. For most of its length, the basilar membrane reacts to incident sound as if the membrane is divided into a sequence of so-called critical bands, each about one-third of an octave (which is about four semitones) apart. This affects many aspects of sound perception: if two tones of markedly different intensity are heard simultaneously and lie in the same critical band, then the louder one will mask the quieter. Two sounds that stimulate different bands, on the other hand, can both be heard, even if one is much quieter than the other. If the sounds heard within the same band are similar in strength, the ear cannot distinguish them and dissonance results. The frequency differences of such sounds correspond to the production of 30–40 beats per second, just as Helmholtz found. Critical bands also help to define scale of notes in music, and to define the spacing of measurement frequencies in sound level meters that are intended to mimic human hearing

Because musical notes are quite rich (that is, not much like sine waves—if they were, then all musical instruments playing the same note would be indistinguishable[31])—that means that they contain many partials, some pairs of which are almost certain to beat at 30–40 per second. The more that do so, the more dissonant a complex sound becomes—hence, critical bands can help explain the *degree* of dissonance. As chords have even more partials than musical notes, it is even more likely that playing more than one at once will produce many beats—so it is more difficult to find a pair of chords that sound pleasant when played together than it is to find a pair of acceptable musical tones. This is not the whole story, though, as there are still cultural differences to account for, and musical context is also relevant.

Modern understanding of the basilar membrane also makes it clear that it is not quite the nervous piano that Helmholtz believed: though there are many individual fibres in the basilar membrane, they are interconnected in such a way that it is impossible for one to vibrate without setting its neighbours in motion too. Nevertheless, Helmholtz's work was correct in many other respects, and, equally important, it was instrumental in increasing the interest in the scientific analyses of sound, hearing, music—and noise.

Vulgar noise

Writing in about 1813, Jane Austen puts these thoughts in the mind of her heroine, Fanny Price, on her return to her sadly vulgar family from the more refined establishment of Mansfield Park, where she has lived for many years:

> she could think of nothing but Mansfield, its beloved inmates, its happy ways. Every thing where she now was was in full contrast to it. The elegance, propriety, regularity, harmony—and perhaps, above all, the peace and tranquillity of Mansfield, were brought to her remembrance every hour of the day, by the prevalence of every thing opposite to them *here*.
>
> The living in incessant noise was to a frame and a temper delicate and nervous like Fanny's an evil which no super-added elegance or harmony could have entirely atoned for. It was the greatest misery of all. At Mansfield, no sound of contention, no raised voice, no abrupt bursts, no tread of violence was ever heard; all proceeded in a regular course of cheerful orderliness; every body had their due importance; every body's feelings were consulted... Here every body was noisy, every voice was loud... Whatever was wanted, was haloo'd for, and the servants halloo'd out their excuses from the kitchen. The doors were in constant banging, the stairs were never at rest, nothing was

> done without a clatter, nobody sat still, and nobody could command attention when they spoke.[32]

So, to Austen and people like her, noisiness was a sure guide to vulgarity and most distressing to someone of more refined sensibilities.

Through the first half of the nineteenth century, quietness became steadily more precious as a mark of taste. This was all the more vital as the increasing social mobility of some people made it imperative that they knew how to behave once they got to their newfound elevations, just in case people doubted whether they really were as posh as their wealth allowed them to hope. While Fanny Price learnt her manners through her upbringing, those whose prosperity was of more recent acquisition needed help—leading to the publication of a great many guides to etiquette in this period, peaking in the 1870s, with more than one guide published per year on average, in the USA and UK alike. People wanted to learn how to behave properly, they wanted others to do the same, and good-mannered behaviour was quiet behaviour: more than ever before, the idea of quietness was bound up with that of taste, gentility, and restraint. Hence in the nineteenth century one of the highest virtues of a domestic interior was its quietness. Ideally, the only sounds would be those that were specifically selected; whether from a pianoforte or reading aloud. The increasing proportion of home-owners at the time may have enhanced the sense of interior spaces as areas where the owner should be able exercise complete control—including over the sounds heard there. Children should be seen but not heard, and woe betide a servant who slammed a door.

Noise was to be controlled in other spaces too: the audiences of musical performances were now increasingly expected to be quiet—their conversations during the performance now being regarded as noise, rather than the music as background—and eventually to confine their noisy applause to the end of the performance only, not between movements. On the other hand, people certainly were expected to make noise at the end of a performance—even if they hated the pieces. If they did, then their feelings could be politely shown only by the brevity of their applause.[33]

But not everyone thought of noise merely as vulgar: the view that it was a real health problem was at last being taken more seriously by people whose views were listened to. Among them was Florence Nightingale, whose *Notes on Nursing* was an 1860 bestseller. Describing noise presciently as an 'absence of care' in a nursing environment, Nightingale makes clear that it is not the level of noise that is the problem but its nature. Whispering, rustling clothes, the 'fidget' of crinolines all lead to 'the horror of a patient, though perhaps he does not know why'. Equally, the startling aspect of sounds was to be avoided: 'every noise a patient cannot *see* partakes of the character of suddenness to him.'[34] Her solution was, as so frequently, zoning: the creation of an area of quiet around the patient.

Unfortunately, patients had to suffer noise from outside the hospital as well as from within, since many new hospitals were constructed in noisy areas, and the quieter districts in which older ones were situated were often becoming increasingly noisy because of the increase in road traffic. Though restrictions of traffic speed near hospitals were attempted in some cases, and straw laid on the streets near others, the effects of such mitigations like these were very limited, particularly because of the appearance of a new phenomenon: the rush hour. Physician Sir Thomas Moore described London street noise in 1869 like this: 'Most of the streets were paved with granite sets and on them the wagons and iron-tired wheels made a din that prevented conversation while they passed by. The roar of London by day was almost terrible—a never varying deep rumble that made a background to all other sounds.'[35]

In many areas of London, street noise was simply too loud to be evaded by withdrawing indoors and playing the piano. One might draw the curtains to avoid seeing the vulgarity outside, but still it entered through the ear. Nowadays, of course, significant acoustic isolation can be achieved through appropriate building construction, primarily by double glazing, sealing gaps, and using sound-absorbing materials. While the dense jumble of artefacts that the Victorians thought of as suitable furnishings, along with deep carpets and thick curtains, helped considerably in the latter respect, any kind of sealing of apertures would have been impossible: houses were heated and lit by combustion, so a ready supply of inflowing air was essential.

Meanwhile, in New York, massive population growth (from 312,710 in 1840 to 813,669 in 1860[36]), together with a network of slow, lumbering streetcars (and occasional herds of escaped pigs or cows), led to major congestion problems and a great deal of street noise.[37]

In Paris, in the 1860s, traffic was not the only source of noise: numerous letters appeared in the French press complaining that the ringing of bells in the early morning was waking people up. One reason for this sudden rise may be the recent spread of effective street-lighting through Paris and other large French cities, which led to later hours being kept.

8

THE FIRST CHAMPIONS OF SILENCE

Considering the growth of noise and of the annoyance caused by it, together with the rise of new sciences, it is perhaps surprising that the noise problem in Victorian London was not tackled directly—by regulating street traffic or muffling mechanical noise sources, for instance. But then, for the Victorians, technology was a triumph. As Sir John Clapham wrote of the 1850s, by then 'Britain had turned her face towards the new industry—the wheels of iron and the shriek of escaping steam. In them lay for the future not only her power and wealth but her very existence.'[1] So, it would not do to criticize modern technology. It would almost be unBritish, as unBritish as . . . street music.

In the nineteenth century, London was by no means unique in being plagued by street musicians: street organs had been illegal in Paris since 1816, though they continued to be heard, nevertheless. In New Orleans, the city council passed an anti-noise ordinance in 1856, making it illegal 'to beat a drum, or blow a horn, or sound a trumpet in any street or public place within the limits of the city'. And in New York, the mayor and the Board of Alderman passed an ordinance against all forms of street music in 1890.

But it was in London that the battle between street musicians and their opponents became a matter of such public interest that many newspapers followed its every turn, largely perhaps because of the personalities and fame of those involved in it.

Though there had been a wide range of street musicians in London for many years, it was in the 1860s that the first major campaign was launched against them. In many ways, they were an ideal target—they were readily identifiable, they were not intrinsic to the functioning of the city, and, perhaps most satisfactorily of all, they were foreign—in fact it was said that there were *no* indigenous street musicians, and, though this was no doubt an exaggeration, it is undoubtedly true that most of them were from Ireland, Italy, and—less often—Switzerland. Often, their nationalities were lumped together under the heading 'Savoyard', though in fact they had long ceased to come from the Italian/French border area of Savoy.

The idea of street musicians as foreign infiltrators was popularized in the press, and in particular by the satirical publication *Punch*, and there were growing calls for laws to be passed against them. Of the many instruments deployed—seemingly as weapons against respectability—it was the organ that was most loathed, as this 1852 description makes clear:

> The piercing notes of a score of shrill fifes, the squall of as many clarions, the hoarse bray of a legion of tin trumpets, the angry and fitful snort of a brigade of rugged bassoons, the unintermitting rattle of a dozen or more deafening drums, the clang of bells firing in peals, the boom of gongs, with the sepulchral roar of some unknown contrivance for bass, so deep that you might almost count the vibrations of each note—these are a few of the components of the horse-and-cart organ, the sum total of which it is impossible to add up.[2]

And indeed it is quite clear that a great deal of street music was a genuine problem and that some performers really used their noise as a weapon, playing loudly outside expensive dwellings until they were paid to go away. Some might feel the same about carol singers today.

It was in 1864 that matters came to a head, with two publications: *Street Music in the Metropolis* by Derby MP Michael Bass, and *Chapter on Street Nuisances* by scientist Charles Babbage. Both marked the culminations of long campaigns by their authors.

Babbage in particular did everything he could to oppose the noises of the streets, using his considerable resources of intelligence, political contacts, and obstinacy. History has on the whole not been kind to him. He is remembered with respect as the originator of the computer, but most of his other work has been either neglected or ridiculed. Certainly, Babbage was an eccentric in many ways, and an obsessive man too. A major reason for the failure of the craftsmen he employed to construct a working mechanical computer was his arguments with them, rather than the limitations of the skills of his time—that the machine was a technical possibility was confirmed in 1996 when the London Science Museum succeeded in building a working version, based on the tolerances obtainable by Babbage's craftsmen.

Babbage attacked noise on many fronts, making numerous court appearances and, like any good naturalist, collecting data to support his case, including his detailed list of 165 interruptions that he suffered over 80 days and his estimate that noise had reduced his working output by a quarter.

Babbage's efforts might have been more successful had he not insisted in characterizing the battle against noise as the battle of the 'intellectual worker' against 'those whose minds are entirely unoccupied'. He includes in his pamphlet a list of 'Encouragers of Street Music':

> tavern-keepers, public houses, gin-shops, beer-shops, coffee-shops, servants, children, visitors from the country, and, finally and occasionally, ladies of doubtful virtue . . .[3]

And he also lists 'Instruments of torture permitted by the government to be in daily and nightly use in the streets of London', comprising

> organs, bass bands, fiddles, harps, harpsichord, hurdy-gurdies, flageolets, drums, bagpipes, accordions, halfpenny whistles, tom-toms trumpets, and, the human voice, shouting out objects for sale.[4]

Given Babbage's love of collecting data (another project involved first counting the number of broken windows in a factory and then investigating the causes of all 164 of them) and obsessive attention to detail, one may assume that this is an accurate list of the more noticeable London street noises of the 1860s.

Babbage's confrontational tactics regarding local noise-makers and in particular his numerous letters to *The Times* met with equally devastating responses from his targets: his neighbours hired musicians to play outside his windows, sometimes using damaged wind instruments to add to the annoyance. Another neighbour blew a tin whistle from the window facing Babbage's house for half an hour every day for several months. A brass band played outside his house for five hours. And, when Babbage left his house,

> the crowd of young children, urged on by their parents, and backed at a judicious distance by a set of vagabonds, forms quite a noisy mob, following me as I pass along, and shouting out rather uncomplimentary epithets. When I turn round and survey my illustrious tail it stops...the instant I turn, the shouting and the abuse are resumed, and the mob again follow at a respectful distance...In one case there were certainly above a hundred persons, consisting of men, women, and boys, with multitudes of young children who followed me through the streets before I could find a policeman.[5]

Meanwhile, Michael Bass adopted a more legalistic approach. Starting in 1863, he made numerous attempts to steer his *Act for the Better Regulation of Street Music in the Metropolis* through parliament. The main thrust of the Act would be to remove the requirement that householders needed to show reasonable cause before the police would respond to their demands to move on street musicians who were disturbing them.

To gather support for his Act, Bass published *Street Music in the Metropolis*. It is a measured and well-authenticated document and very far from the tirade it was lampooned as at the time. It includes accounts from clergymen, lawyers, academics, teachers, composers, and musicians. But the greatest emphasis is on writers and artists, and a letter written by the most famous of their number, Charles Dickens, is the centrepiece of the book. Dickens writes that he and his cosignatories 'are daily interrupted, harassed, worried, wearied, driven nearly mad, by street musicians'. Writing of 'brazen performers on brazen instruments', he adds: 'No sooner does it become known to those producers of horrible sounds that any of your correspondents have particular need of quiet in their own houses, than the said houses are beleaguered by discordant hosts seeking to be bought off.'

And, the twenty-eight-strong list of signatories was impressive, and included Wilkie Collins, E. M. Forster, William Holman Hunt, John Everett Millais, Alfred Lord Tennyson, and (not very surprisingly) Thomas Carlyle.

The publications of Babbage and Bass won the day. Later in the same year that they appeared, Bass's Act passed into law. It was the beginning of the end for street musicians, though that end was slow in coming—so slow that, when Babbage was on his deathbed, an organ grinder played outside.

Electrifying sound

By the 1870s, the new noises of industry were to be heard in cities throughout much of the world. This is how August Strindberg describes the scene in Stockholm:

> Far below him rose the clamour of the newly awakened town; down in the harbour the steam cranes whirred, the bars rattled in the iron-weighing machine, the lock-keepers' whistle shrilled, the steamers at the quayside steamed; the Kungsback omnibuses rattled over the cobblestones; hue and cry in the fish market, sails and flags fluttering in the water, screams of seagulls, bugle-calls Skeppsholm, military commands from Sodermalmstorg. Workmen in wooden shoes clattered down Glasbruksgatan, and all this gave an impression of life and movement.[6]

As noise spread around the world, it also spread through the day, until, in many places, there were no quiet hours left. So, with a lack of quiet blocks of time, there was increased pressure to establish quiet blocks of space: zoning was applied more and more widely, until many large towns and cities had very well-defined industrial, commercial, and residential zones with almost no overlap of their functions. But, in its wake, this solution generated the problem that was to dominate concerns from then on: transport noise. Most people could no longer work from home, nor walk to work. For them to get from their relatively quiet home zones to their bustling work zones, transport of some kind was essential. And a lot of it made a terrible racket. In New York City the construction of the first elevated railways began in earnest. They were soon to lead to the city being widely regarded as the noisiest in the world. As the decades passed and people became more accustomed to the idea of commuting, some moved back into the countryside again, establishing dormitory towns linked to the centres by fast and noisy transport links.

This problem and a proposed solution are described in Sir Benjamin Ward Richardson's 1875 description of a healthy utopia called, rather unimaginatively, *Hygeia*. Zoning, together with sound insulation, is the principle on which the whole city is constructed:

> Beneath each of the walkways is a subway, a railway along which the heavy traffic of the city is carried on... The streets of the city are paved throughout in this same material. As yet wood pavement set in asphalt has been found the best... The subways relieve the heavy traffic, and the factories are all at short distances from the town, except those in which the work that is carried on is silent and free from nuisance.[7]

Though such zoned cities were commonplace within decades, the sound insulation technology was—and is—not.

In any case, if quiet cities were really what were wanted in the late nineteenth century, people were not going about getting them in a very sensible way: no sooner had modern buildings been built, using modern building methods that would allow some noise to be kept out, than their inhabitants proceeded to fill those interior spaces with newly available noise-making technology. In 1877 Thomas Edison's phonograph was launched and began to add its contribution to the din. And the word 'microphone' was transferred from its dodgy first incarnation to its world-changing second version, invented by Arthur Hughes in 1878.

It was in the 1860s that Alexander Graham Bell emerged as the official inventor of the telephone—whether this is an honour that he really deserved is a story for a whole book, but he certainly owned the money-spinning patent. However, he was not apparently best pleased with the masterpiece to which he struggled so hard to maintain the rights, refusing to have one in his own home and downplaying its significance in comparison to his researches in heavier-than-air flight and his invention of the photophone, a device that transmitted sound by means of light (see Fig. 18). The photophone was based on selenium, a material whose electrical resistance varies with the amount of light falling on it. The light in question was a beam reflected by a thin mirror that vibrated when spoken at, and the selenium controlled the flow of electricity to a telephone receiver.

While their inventor remained unimpressed by telephones, many others were much more interested, with Queen Victoria herself waxing enthusiastic about them by the 1870s.

And they needed some enthusiasm. Both their persistent ringing and all the shouting that was required to use them helped fracture further the inner peace and harmony that were still fondly imagined as characterizing domestic spaces.

The fault lay not just with poor amplification and intrusive background noise but also with the very limited bandwidth of these early devices. However, despite such practical shortcomings, new electro-acoustic technologies took hold of the Victorian imagination: attempts were even made to transmit live music by

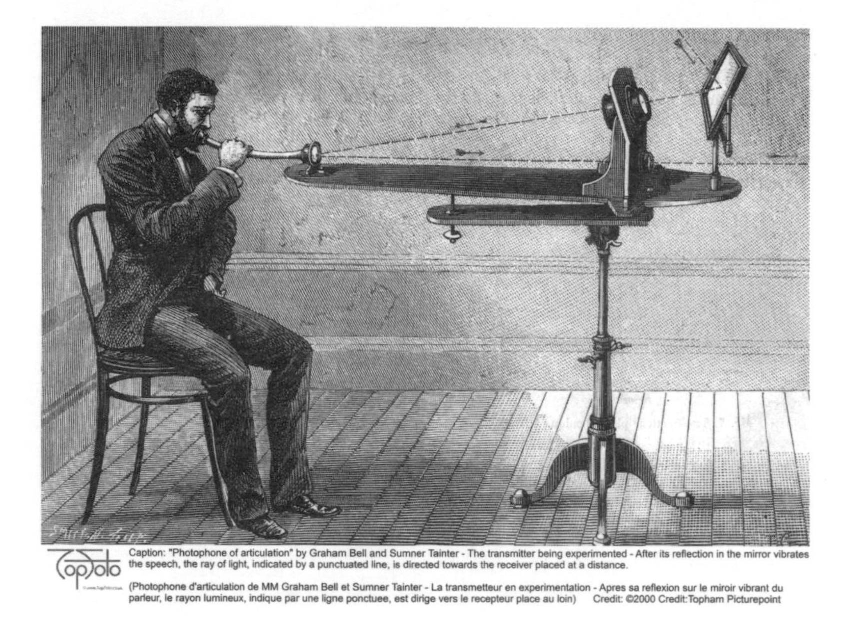

FIGURE 18. Bell's photophone. The user's voice causes a thin mirror to flex, so that the brightness of a light beam reflected from it to a selenium cell varies. The electrical resistance of the selenium varies as a function of the amount of light incident on it, and this is used to modulate an electrical signal.

TopFoto.

telephone, an ambitious project that would not really work even with today's devices.

New knowledge

The late nineteenth century was also the time of the greatest acoustician that no one has heard of: John William Strutt, slightly better known as Baron Rayleigh. His *Theory of Sound* (1877–96) covers the whole science of acoustics as it was known at the time and provides new insights and discoveries in many of its facets.[8] Among other important contributions, Rayleigh explained correctly how humans and animals are able to determine the direction from which a sound is coming. For high frequencies, where the sound wavelength is small compared to the distance between the ears, the difference in sound levels at each ear is used, while, for lower frequencies, it is the time between arrivals at the ears that is key. Unfortunately, these two effects do not overlap much for human ears, so sounds in the 1 kHz–4 kHz region are difficult to locate, which is why it is so tricky to track some species of cricket and grasshopper.[9]

Theory of Sound even includes a mathematical account of ultrasound at a time when the only effective artificial source of it was a whistle and it was, for all practical purposes, undetectable.

On 29 March 1871, noise control suddenly became of urgent interest, if only in one part of London. On that day, with a great deal of pomp and razzmatazz, the Royal Albert Hall was opened—though not, as it turned out, by Queen Victoria, as had been planned. When the moment came, she was said to be so overcome by the occasion that she could not bring herself to speak. So it was the Prince of Wales who did the honours, announcing rather oddly: 'The Queen declares this Hall is now open.' Unfortunately, Victoria was not the only one not to be heard there: as soon as the inaugural concert began, the poor acoustic design of the interior became disturbingly obvious, due to a very disconcerting echo. No one was impressed. The equally unimpressive state of architectural acoustics at the time is readily apparent from the fact that, despite the prestige of the site and the plentiful funding available, the numerous attempts to improve the Hall in the next few decades all came to nothing or actually made things worse. It was nearly a century later (in 1969) that the introduction of mushroom-shaped acoustic diffusers finally resolved the problem (see Fig. 19).[10]

A similar post hoc approach to noise problems was being adopted across the Atlantic at about the same time. New York City's elevated railway system (see Fig. 20), known as the 'L', was notorious for the level, character, and spread of its noise; so much so that in 1878 the

FIGURE 19. The original interior of the Royal Albert Hall (a), and a contemporary view (b), with diffusers in place.

(a) Heritage Images; (b) Graham Salter/Lebrecht Music and Arts.

A WEEKLY JOURNAL OF PRACTICAL INFORMATION, ART, SCIENCE, MECHANICS, CHEMISTRY, AND MANUFACTURES.

Vol. XLI.—No. 17. [NEW SERIES.] NEW YORK, OCTOBER 25, 1879. [$3.20 per Annum. [POSTAGE PREPAID.]

THE PROGRESS OF ELEVATED RAILWAYS.

The ancient story of the intruding camel, who begged a shelter for his head in his master's tent and ultimately crowded in his unshapely body, to his master's great discomfiture, is paralleled in the history of elevated railways in this city.

The main reason for the adoption of this form of rapid transit was the cheapness with which it could be supplied. The camel's head was not attractive, but it was easily let in, and promised an easy removal should such an issue prove desirable. Fig. 1, page 258, shows what an early form of the original West Side elevated road was like; not the earliest form, however, for that was of considerably lighter construction. The large engraving presented below gives a hint of the enormous possibilities of the structure which has taken possession of so much of the city. As a specimen of bold, clever, and original engineering it is admirable. Its effect upon the fine avenue it overshadows is quite another matter. So, too, is its probable influence upon the region it traverses as a site for dwellings. The utter inadequacy of any cheap structure of slight capacity (such as the elevated roads were at the start) to meet the wants of a city like New York, and the fallacy of the assumption that such a rapid transit road was advisable on the score of economy, were repeatedly enlarged upon by the SCIENTIFIC AMERICAN in the early days of the system; and the result has more than justified the position then taken. During the past five years, indeed during the past three years, the system has expanded from four or five miles of roadway of the lightest description, supported by single posts, to ten times as many miles of massive and costly structure already in operation, and nearly twenty miles more approaching completion—structures which almost monopolize four of our principal avenues and large portions of several down-town streets, and represent an investment of $13,000,000.

The system which has attained such stupendous results began in an extremely modest way in 1868, and for several years it was represented by half a mile of experimental road on Greenwich street. The New York Elevated Railway Company was organized in 1872, and during the summer of 1873 the road slowly crept up Greenwich street and Ninth avenue as far as 30th street. In 1876 it extended from the Battery to 61st street, and during the succeeding years it was further extended to Central Park, and to a considerable extent was made a double track. Though the new road was heavier than the parts of the line first constructed, the system of single supports was adhered to, and the general character of the road was sustained. During the early part of the current year the track was extended to 83d street, and the original track on Greenwich street has recently been replaced by the heavier structure of the later road.

In 1878 the Gilbert, afterwards known as the Metropolitan road, was completed to 59th street—a double track occupying the whole of the narrower streets down town and the middle of the wide Sixth avenue, and surpassing in solidity and cost anything previously dreamed of in the way of high level road making. The cost of constructing and equipping the five miles from Morris street to 59th street, with half a mile of road from Sixth avenue to Ninth avenue, through 53d street, was officially reported in March last as $10,300,000.

During the same year the New York Elevated Railway

[Continued on page 258.]

THE ELEVATED RAILWAY AT 110TH STREET AND EIGHTH AVENUE, NEW YORK CITY.

FIGURE 20. The Elevated Railway at 110th Street and Eighth Avenue, New York City, in 1879.

Scientific American, 25 Oct. 1879 (cover); reprinted with permission.

railway company called in Thomas Edison to analyse the source of the noise. Equipped with a phonautograph (see below), the inventor obliged. Sadly, these first ever environmental sound recordings by the first ever hired acoustic consultant have not survived.

Edison was not asked to solve the noise problem: that was accomplished by Mary Walton, an environmentalist before the term existed, who had cut her teeth on removing some of the noxious components of the belching smokestacks of American factories. In 1879, her campaigning to quieten the L by installing asphalt-topped cotton- and sand-filled boxes over the rails was successful, and the noise improvement that resulted was significant.

It was at about this time too that many factories started to try to quieten their noisy machinery. Though treatments were applied in a hit-and-miss and uncoordinated fashion, they at least gave some degree of protection for the long-suffering workers. This relief was very much a lucky side-effect, however: the main motivation for the quietening of machinery was to reduce wear and the concomitant financial losses.

It is noticeable that, in the list of signatories of Bass's anti-street noise campaign, there are no medical doctors. But, a generation later, one trade was causing hearing problems so severe and so notorious that it gave rise to the name of a medical condition: boilermaker's ear. The ailment received such attention that, in 1886, it finally led to an actual scientific investigation. It was the first quantitative study in the field ever carried out (and the last until the 1920s). The investigation was conducted by Thomas Barr, a surgeon who worked for Glasgow Ear Hospital, and whose experience of the stream of Glaswegian workers suffering from severe hearing problems had prompted him at last to study the causes at first hand.

Several related jobs were involved in boilermaking, and the worst of these was that of the 'holder-on', whose responsibility it was to stand inside the boiler holding a large hammer, the end of which he had to press against the inner end of a rivet. As he did so, the outer end of the rivet was struck by a hammer. As Barr described it, 'the iron on which they stand is vibrating intensely under the blows of perhaps twenty hammers wielded by twenty powerful men'.[11] When Barr himself entered such a boiler, he described it like this:

> we are conscious not merely of the sound waves, like blows, producing their terrible effects upon our ears, exciting therein sharp, painful, intolerable sensations, but our bodies seem to be enveloped in invisible, yet tangible waves which we actually feel striking against our heads and our hands... After such an experience one is surprised that the delicate mechanism in the interior of the ears can retain its integrity for a single day under the action of these blows.[12]

Barr was unusual, not only in voluntarily subjecting himself to the conditions his patients faced, but in attempting to measure them. As such he was the first to tackle a problem that persisted for many decades and is still incompletely resolved: how to relate the subjective features of noise to the objective ones. There was almost no equipment that he could use to make objective measurements of any kind and all he could try was to take a not-very-portable phonograph with him into the interior of the flue, the area of the boiler in which the sound was most intense. While 'riveters and caulkers were hard at work without and within', he attempted to make some recordings. Perhaps not surprisingly, the results were none too clear, but he was at least able to conclude that 'the indentations [on the wax cylinder] were small and closely arranged, indicating the great height of pitch of the notes and contrasting with the large widely separated indentations caused by the human voice'.[13]

Undaunted, Barr proceeded to examine the hearing of 100 full-time boilermakers, whose ages ranged from 17 to 67. Again, he was at pains to make his work as objective as possible, and his simple and fairly effective approach was to measure the distance from the ears of the men at which they could just hear the ticking of a watch. In order to make a comparison with normal hearing, Barr tried the same approach with people who, as far as he could judge, had no abnormal hearing loss. He found that the average distance for those men was 36 inches. By comparison, the average for the boilermakers was 3.6 inches. Half the men could hear nothing at all unless the watch was actually pressed against the ear (in which case, the sound waves were transferred to the inner ear by conduction through the mastoid bone, missing out the eardrum entirely)—and many of them could not hear it even then.

Noting that the ticking of a watch is sometimes easier and sometimes more difficult to hear than a voice, Barr carried out experiments, less objective but still controlled, in which he whispered, spoke normally, and finally spoke loudly a number of words at a distance of a yard from the men's ears. Again, the results were striking, with 13 of the men unable to hear even loudly spoken words at one yard's distance.

Finally, Barr asked the men whether they experienced any difficulties in hearing public speakers—54 said they had such difficulties, and 24 that they could not hear enough to understand anything.

Although in his search for objectivity Barr is far ahead of his time, the concluding part of his paper is very much a product of the mindset of his day. His main concern is the impact of the workers' hearing loss on their 'social comfort and usefulness'—in particular, their ability to hear public speakers. Of particular concern to Barr was that the men were put off attending church services. So, perhaps his proposed solution to the problem of boilermakers' ear should not surprise us too much: 'this serious defect might be met by providing them with small spaces of worship, having good acoustic qualities, and supplied with speakers possessed of a strong voice and clear articulation.' He also urged clergyman and other public speakers 'to eschew, if possible, moustache and beard', so that some degree of lip-reading was possible. As he pointed out: 'Deaf ladies who wish to conceal their infirmity shun the society of moustached and bearded men.'[14] (Gentlemen were supposedly more relaxed about being hard of hearing.)

If Alexander Graham Bell or Lord Rayleigh ever read Barr's paper, perhaps they were a little disappointed that he had had recourse to a phonograph rather than to their own inventions for capturing sound: by this time, Bell, had developed another innovative way to convert sound to visible traces, albeit a rather unnerving one. In 1857 Édouard-Léon Scott de Martinville had invented the phonautograph, the world's first device that could make airborne sounds visible, by using the vibrations produced by it to inscribe patterns on moving strips of soot-covered paper or glass. But constructing suitable diaphragms for it was problematic. Bell's solution was to use a ready-made one: a dead person's ear was pickled in oil to keep it supple and a straw was attached to its drum. This straw was then set up so that it scratched the surface of a blackened strip of paper, which was moved slowly past, so that a trace on the paper was recorded. Despite being a surgeon, and so presumably well supplied with dead people's ears, Barr doesn't seem to have considered using this 'ear phonautograph', as it was called.

Even if he had, he could not have got more than a pictorial representation of a sound from it. But a little later, in 1880, Lord Rayleigh had developed something truly ground-breaking: an instrument objectively to measure the power of sound, later called the Rayleigh disc. The device is a simple one: just a light diaphragm supported by a fine wire. Its principle was the same as Bell's photophone, but the effect on the disc was rather different: rather than trembling in response to the sound waves, it was rotated by them, through an angle that was dependent on their intensity. A light beam was reflected from the disc, allowing the deflection angle to be accurately measured.

The disc was to be used for decades in measurement laboratories as a way of monitoring sound pressures. However, it had considerable practical shortcomings: in particular, the slightest breeze would disturb it. Consequently, it was very difficult to use and had to be encased in a curtained enclosure, like a magician's trick. But nevertheless, as the first ever means of making an objective measurement of sound, it was a significant step forward.

FIGURE 21. A contemporary illustration of the 1883 Krakatoa eruption.

Classic Image/Alamy.

Memorable sound

At 10.20 on 27 August 1883, on a tiny island just off the coast of Jakarta, in what is now Indonesia, the loudest noise in recorded history was generated when the volcanic island of Krakatoa exploded, releasing energy equivalent to about 13,000 times that produced by the bomb that was dropped on Hiroshima (see Fig. 21). It could be heard clearly more than 4,000 kilometres away. The sound travelled around the world many times, and could still be detected by measuring instruments five days later. Thirty-metre-high tsunamis swept the Indian and Pacific oceans, and smaller versions appeared in the English Channel—though they arrived so soon after the explosion that they can only have been caused by airborne shock waves radiating from the island. According to R. M. Ballantyne, in his 1889 novel *Blown to Bits*, 'the *world* heard that crash. Hundreds, ay, thousands of miles did the sound of the mighty upheaval pass over land and sea to startle, more or less, the nations of the earth.'[15]

On acoustical topics as on others, the turn of the twentieth century was characterized as much by nostalgic reminiscences of a vanishing past as by anticipations of an exciting future. While novelists like Thomas Hardy frequently used descriptions of rural soundscapes to capture the feeling of an unspoilt past, often underplaying how dull and unpleasant it must frequently have been to live in one, other writers and commentators were doing the same with urban spaces, and fondly describing the times when streets of London were full of the cries of street vendors and the music of street musicians—despite the fact that it was only a few decades since the latter had been widely viewed as the bane of modern city life.

In a similarly retrospective tone, the anthropologist Karl Bücher lamented the days when machine noise was comfortingly rhythmic. Based on the idea that dancing to rhythmic sounds is a universal human tendency, Bücher moved on to the rather more contentious conclusion that the reason music developed was specifically to synchronize communal work activities and to encourage recalcitrant individuals to take part. Consequently, the most frustrating sounds (for everyone, but especially for workers) were those where rhythms were present but, tantalizingly, undanceable-to—a state of affairs that characterized practically all modern factory machinery. Looking forward to the new century, Bücher hoped that technology and art might somehow achieve a higher rhythmical unity.[16]

The world's first Society for the Suppression of Noise, formed in London in the 1890s, had a similarly nostalgic turn of mind, focusing mainly on wiping out such shocking modern ills as the motor car—despite the fact that only a few dozen of the machines were on the road at the time and can hardly have constituted a significant noise problem, especially as they tended to break down every few kilometres. Despite the rarity of the sound of the motor horn, laws were passed against its use. Perhaps it was disappointment that exacerbated their irritation: in the 1890s, in London and New York alike, what people liked about the idea of the horseless carriage was that it would be quieter than one with a horse attached. The *Scientific American* said at the time: 'The noise and clatter which make conversation almost impossible on many streets of New York at the present times will be done away with, for horseless vehicles of all kinds are always noiseless or nearly so.'[17]

New worlds of noise

A newfound concern with environmental issues, including noise, was now evident in the urban United States in general and New York City in particular. This new focus seems to coincide with the Panic of 1893; almost forgotten now, it was the worst depression that the USA had experienced, and it was caused by the collapse of enormously overpriced railway stocks. Perhaps again it was the sudden disillusionment with railways that put people in the frame of mind to focus on their negative aspects—including noise.[18]

By 1900, most European countries had regulations against night noise disturbances. In England, these prohibitions extended to trade-based noise, but such sounds were excluded from the legislation in France, Holland, and Germany at this time (though conversely in Germany other types of noise were excluded by day as well as by night). While these regulations probably did reduce the problems of traffic noise, nothing similar was enacted to protect factory-workers, who were still going as deaf as ever.

Out of the reach of such workers was a whole new range of devices for those with both hearing problems and money. As an 1895 catalogue of *Otological Instruments to Aid the Deaf* explained:

> Sensitive persons, particularly ladies, have an aversion to advertising their affliction in public by the use of many of the usual forms of hearing instruments. To meet this very natural objection, such instruments have been ingeniously combined with fans, parasols, umbrellas, muffs, handbags or reticules, bouquet holders, opera

> glasses, &c. Other instruments are attached to the head and ears, and may be concealed by the cap, hat, bonnet or hair... For gentlemen, walking sticks and umbrellas of various sizes have powerful sound collectors fitted to them; also dinner plate holders and field glasses and the inside of the ordinary silk hat.[19]

Just a few years later, in 1906, an even more novel type of hearing aid was developed, especially for those of an outdoors nature. Virginia Hollyday's invention was a combined speaking tube, hearing trumpet, and garden seat. Once seated on it, couples could converse undisturbed by ambient noise, though, perhaps in obedience to some bizarre Edwardian concept of modesty, the speakers had to sit back to back to use it.

Meanwhile, the phonograph had developed from a barely useable, low-quality curiosity in the 1870s to a well-made and quite effective sound recorder and player. It was so popular by now that the whole phenomenon of pop hits was beginning to take hold. Though the quality was still terrible by modern standards, the idea of preserving sounds—and especially voices—struck the Victorians as a very attractive one, obsessed as they were by the idea of transcending death, through their great pompous funerals and love of spiritualism and ghost stories. Alfred, Lord Tennyson, was particularly enthusiastic about the new technology, and his voice is one of the earliest that survives.

At the end of the century, a new scientific breakthrough was to have profound implications for the control and use of sound: Karl Rudolph Koenig, a German physicist, was fascinated by recent reports that the human ear could not hear tuning forks of pitches higher than around 30 kHz.[20] He had already (in the 1870s) constructed large sets of tuning forks to investigate the effects of simultaneous audible tones. In 1876 his 'grand tonometer' was demonstrated at the Philadelphia Exposition—it contained 692 forks, ranging from 16 Hz to 4096 Hz.

Now, by making tuning forks as small as 1 centimetre in length, with 1 millimetre gaps between their prongs, Koenig was able to make sounds far too high for anyone to hear. But how was he to know that such forks actually made sounds at all? And, if they did, how to assign a frequency to them? To investigate this he used a device now called a Kundt's tube. The tube is a thin-walled glass cylinder containing fine light powder, laid on its side, the effective length of which can be altered by a piston. When the tube is touched by a vibrating tuning fork generating sound waves that are exactly twice the length of the tube, or fractions of that length, standing waves are set up in the tube, and the powder arranges itself in repeating patterns, with piles of dust at the motionless points (nodes) of the standing waves. The nodes are a half-wavelength apart.

Inspired by Koenig's work, English physician Sir Francis Galton investigated this new type of sound further in the 1880s: instead of minuscule tuning forks, he constructed equally tiny (but much more powerful) whistles. Galton decided to find out how the hearing of animals compared with that of humans and so he selected numbers of individuals from different species. First, he patiently held the whistle close to their ears—without blowing it—until they were no longer bothered by its presence. Then he blew the whistle and noted whether the animal pricked its ears. Unfortunately, having carried out this ground-breaking experiment, Galton rather devalues it by not giving his quantitative results. His practical conclusions are pretty much limited to his statement: 'Of all creatures I have found none superior to cats in the power of hearing shrill sounds.'[21] Apparently he did his work mostly to satisfy his own curiosity, and for fun. But fun was not what William Sabine was looking for…

Sabine and the sabin

Sabine was undoubtedly the greatest practical acoustician of the nineteenth century—and perhaps of any other. It is really he who is the father of effective architectural acoustics. His fame, and his mastery of the subject too, were achieved through a single masterwork: in 1895, he tackled the challenges of the Fogg Lecture Hall, the most important space in the Fogg Art Museum, which had proudly opened to the public just a few months before.

The room was dogged by such problems of reverberation that it was hardly useable, and the owners of the museum turned to the eminent Harvard University Physics department for help. Such was the state of architectural acoustics at the time that the senior staff there were privately doubtful that anything could be done—or, as they put it, they were far too busy. Consequently, they passed on the request to the most junior person available, Wallace Sabine. And he was not just junior—he was also young, had no Ph.D., and, rather more relevantly, had no special knowledge of acoustics of any sort.

Perhaps it was his inexperience and freedom from the plethora of misconceptions and half-truths that then characterized architectural acoustics that led to his success. (At the time, some engineers were fond of stretching wires across reverberant spaces to quieten them, while others were decorating them with vases and vaguely hoping the ancient Greeks were right after all.) In any case, what he did was simply to approach the problem in a logical, stepwise manner, which did not rely on acoustical assumptions at all. First, Sabine tracked down a space that was similar to the Fogg in terms of dimensions and function—but that had excellent acoustics. He found it in the Sanders Lecture Theatre. So, *why* was it so superior? The most obvious difference was the fittings of each room, so Sabine proceeded to transfer various moveable items from the Sanders to the Fogg. Assisted by some of his more loyal students, he carried groups of seat cushions from the inconveniently vast

Sanders to the Fogg, and, once each set was in place, he measured the reverberation time at different frequencies. Having no special equipment for the job, he simply used a set of organ pipes, timing how long it took the sound of each pipe to fade to inaudibility. Encouraged by this success, he managed to convince several more students to help out, and seated them in the theatre in differently sized groups while again determining the reverberation time—from which he learned that one person was equivalent to six cushions at most frequencies. Finally, he tried the effects of a range of rugs that varied in area and thickness.

An added challenge was the need to carry out all the work at night and to return the halls to their original state before morning, since the halls were in almost constant use during the day. As it turned out, this was of great benefit, as it meant background noise levels were low throughout his experiment.

Sabine's nightly activities, and the simple but masterful analyses he carried out on them afterwards, led to his establishment of a number of key principles of architectural acoustics, foremost among which was his definition of reverberation time: the duration required for the intensity of the sound to drop from the starting level by an amount of 60 dB. He also derived the key equation of indoor noise control:

$$RT_{60} = 0.049V/Sa$$

where RT_{60} is the reverberation time, V is the volume of the room in cubic feet, and *Sa*, as well as being the first two letters of Sabine, is the total absorption (in which a is the average absorption coefficient of room surfaces and S is its surface area).

A great deal of science in general and acoustics in particular proceeds like this, with centuries—or in this case millennia—in which an effect is vaguely understood and loosely described, but cannot be used or controlled until the mathematical relationship between properly described and measurable quantities has been established. In architectural acoustics, the final stage is to relate the measurable quantities to subjective impressions—which Sabine did simply by measuring the reverberation times of as many spaces as he could that were judged to be acoustically 'good' ones. He found that concert halls had reverberation times of 2–2.25 seconds, while good lecture halls had reverberation times of slightly under 1 second. The Fogg Museum lecture room, on the other hand, had a reverberation time of over 5 seconds. Reverberation time is still the most important parameter for describing the acoustical quality of a room. Ideal values depend on the purpose to which the space is to be put—while a theatre should preferably have a time of ½ a second to 1 second (so that each word can be heard clearly), that of a church used for musical purposes should be between 2 and 4 seconds. Concert halls are more tricky, as the ideal reverberation time depends on the sort of music being played there. For classical composers such as Mozart, about 1.4 seconds is right, while those grand Romanic symphonies are at their sweeping best at around 2.1 seconds. (There are many other criteria that are relevant to the quality of a concert hall, including low background noise, uniformity, and clarity, which depends on the *amount* of reverberation, rather than its time.)[22]

Finally, having transformed the whole of architectural acoustics in a matter of weeks, Sabine returned to his original instructions and used sound-absorbing materials to reduce the reverberation time of the lecture theatre. So successful was the theatre thereafter that Sabine was commissioned as the acoustical consultant for Boston's prestigious Symphony Hall, the first concert hall to be designed using quantitative acoustics. His design was a great success, and the Symphony Hall is generally considered one of the best symphony halls in the world, even today. In addition, the unit of sound absorption, the sabin, was named after him. What happened to the 'e' is a mystery; it has been suggested that there was a concern that people might be a bit squeamish about associations with the Rape of the Sabines—but 'rape' in those days (and particularly regarding the Sabines) usually meant 'kidnapping', so it seems unconvincing. Whether people blushed when Sabine introduced himself to them is unfortunately unknown, as is his reaction to the spelling.

One sabin is a very handy measure, being equal to the sound absorption of 1 square foot of a perfectly absorbing surface such as an open window. In fact, Sabine himself called it the Open Window Unit. Today, 1 square metre of 100 per cent absorbing material has a value of one metric sabin.

Sabine spent the rest of his working life happily making places sound better. Some of them were illustrious ones, like the chamber of the House of Representatives in Washington, DC, the Military Academy chapel at West Point, and the Halifax Cathedral in Nova Scotia, and, in one of the first scientific attempts actually to quieten a workspace, the Remington Typewriter Company.

9

THE NOISE CENTURY

The 1900s was a period of tumultuous change in western Europe and North America alike, as the rate of technological change increased and the sizes of cities swelled. Suddenly, with the new century, new technology was in the air. Just a few years after the Wright brothers clattered falteringly aloft, crowds flocked to see Louis Blériot complete his triumphant flight across the English Channel. And in the USA, Henry Ford began his revolution of the modern world with the launch of his new style of factory: based on production lines and able to produce cars at more than ten times the rate that had previously been possible. Prices dropped and demand soared, and the sound of the internal combustion engine became more widespread, with a concomitant increase in noise.

Nor were new noise sources confined to the outdoor world. The telephone, invented decades before, now really took hold of the public imagination, and the number of subscribers burgeoned. The turn of the new century marked the death of one of its earliest fans: Queen Victoria, whose funeral was on 2 February 1901—the first year of the new century as it was then regarded. And a small acoustical mystery attends her death. The noise of the cannon fired at the ceremony was heard over wide areas of central London—but not in its outskirts, nor in the countryside immediately outside it. Yet it was heard loudly at a number of villages in an approximate ring around 150 kilometres from the source. The most likely explanation is temperature inversion, in which the usual reduction in temperature of height is reversed owing to the effects of ultraviolet solar radiation, leading to a temperature peak at around 50 kilometres. The progressively cooler air in the lower part of the atmosphere causes the paths of sound waves to be deflected upwards,[1] causing a rapid drop in the loudness of the sound at ground level beyond a few kilometres from the source. But, once the region of increasing temperature is reached, the waves curve down again and return to the Earth at a great distance from the source. Temperature inversions are often produced when a warm front pushes a warm air mass over a lower, cooler one, and can also occur on clear nights when the earth's surface radiates its heat away more rapidly than does the air above it. Mining companies pay great attention to such effects today, and check on weather conditions before commencing blasting operations to ensure that such displaced noises do not trouble people living nearby.

Meanwhile, Edison's phonograph (and Marconi's wireless, though as yet usually without a loudspeaker) continued their colonization of the home. Though new and exciting (and/or annoying) in themselves, these electro-acoustic devices were in fact members of a breed soon to become almost extinct, thanks to the work of American engineer Lee De Forest. In 1906, De Forest constructed the world's first electronic amplifying device, the triode. The basic principles of electronics had been established for some years, but their applications were very limited without the possibility of increasing the strength of a signal. The triode[2] did just that, making really noisy sound reproduction possible at last.

To begin with, valve amplifiers based on the triode were used to boost telephone signals, which may at least have slightly reduced the noise of people shouting down phones, but by the 1930s they were being used to build radios (then called wireless sets) that could be listened to without the need to put an earphone in one ear, a finger in the other, and concentrate very hard.

Of course, compared to today, the numbers of radios, phones, and phonographs were tiny. But, through the newly efficient media system, everyone seemed to know, and to be talking, about them.

General technological growth was very real though: between 1860 and 1914 manufacturing production in the USA increased by a factor of twelve, mainly thanks to the American innovation of interchangeable parts and the use of production lines. Meanwhile, in Europe, to take a random example, Hanoverian trams surged in frequency, with over 850 passing a monitoring point in every 24 hours.

In London and New York in particular, populations were rapidly increasing, and infrastructures designed for very different numbers of people and ways of life struggled to cope. London streets became a piecemeal mass of surfaces, with the classier areas furnished with the latest developments, like tar macadam, though this material remained the rarest of all until well into the 1910s. Meanwhile, the majority of thoroughfares consisted of a random-seeming sequence: cobbles, granite blocks, asphalt, and wood. Of these, cobbles and granite were especially noisy, while, for a time, it seemed that wood was the answer to the noise problem, being quieter than traditional materials and cheaper than asphalt. Wooden road surfaces were popular too in Australia, the United States, and Canada at this time. However, in all areas they were insufficiently hard-wearing to cope with rising traffic densities and had mostly disappeared by the early twentieth century.

The noise of hooves and iron wheels on the streets, plus a number of motor cars, was such that the riders of horse-drawn vehicles were forced to add further to the noise by shouting at each other when they wished to pass. And that meant a lot of shouting: in 1907, Germany had 27,000 cars but over 2,000,000 horses for transport.[3]

In the United States, noise complaints were on the increase: a letter to the *American Magazine* demanded that it should declare war on the 'Noise Devil' as a danger to health, learning, property values, city growth, and human relationships—a list that it would be hard to better today.[4] One well-publicized noise was the rattle and crash of early morning milk deliveries, so much so that some New York milk companies began to fit their horses with rubber shoes and their wagons with rubber tyres, and to use them to offer 'noiseless milk'.[5]

Typically, not everyone thought noise was a problem, and those who did were divided on which kinds were a nuisance: George Trobridge, a landscape painter writing in the *Westminster Review* in 1900, claimed that the noisy world produced 'lunacy and the nervous diseases', mainly because of sleeplessness.[6] Like Babbage before him, Trobridge emphasized in particular the added annoyance of noise made unnecessarily. One of the most irritating sounds that emerged from a nearby bottle works was that made by the night workers who 'amuse themselves by singing and shouting at their work'.[7] On the other hand, the sound of an iron forge—albeit fairly distant—he found to be 'musical' and actually to help him sleep.

This appreciation of the musical qualities of industrial facilities, though not their supposed sleep-inducing qualities, would certainly have found favour with one group of contemporary artists, the futurists.

Futurism, an artistic movement spawned by the Italian artist Fillipo Marinetti, published its manifesto in 1909. It is likely that a major stimulus was the work of Karl Bücher and in particular his hope—published in 1896[8]—for a 'rhythmical unity' of technology and art. The manifesto praised industry and the noise it made and encouraged artists to incorporate it into their works. The idea that an art movement should need a manifesto at all indicates the new fusion that Marinetti had in mind: for him, art had for too long stood aloof from the world. And it was not only the noise of industry that he was keen on: 'We will glorify war!'[9] he excitedly proclaimed. One of the first to respond was Russian composer Arsenij Avraamov. His most famous work, *Simfoniya gudkov*

(*Symphony of Factory Sirens*), requires the use of (recordings of) navy ship sirens and whistles, bus and car horns, factory sirens, cannons, the foghorns of the entire Soviet flotilla in the Caspian Sea, artillery guns, machine guns, 'hydro-airplanes', a specially designed 'whistle main', and—possibly to entice in the more traditional music-lovers—renderings of the *Internationale* and *Marseillaise* by a mass band and choir. The piece was conducted by a team of conductors using flags and, rather alarmingly for the audience, pistols.

But, for Avraamov, this sort of thing was only the start. In his 1916 article 'Upcoming Science of Music and the New Era in the History of Music', he proclaimed the future of music: 'By knowing the way to record the most complex sound textures by means of a phonograph, after analysis of the curve structure of the sound groove, directing the needle of the resonating membrane, one can create synthetically any, even most fantastic sound by making a groove with a proper structure of shape and depth.'[10] While completely wrong about the means, Avraamov was shortly to be proved exactly right about the end result.

Fighting back

Meanwhile, captains of industry, ruling elites, and the medical establishment alike still showed little concern about the rising levels of machinery noise, which continued to blight the lives of workers. When the impact of such noise was mentioned, it was still in the context of inefficiency: 'A large part of the noise in a manufacturing plant may be translated into loss of power, unnecessarily rapid depreciation in equipment and as reduced efficiency of employees resulting from the distraction which is created and from the indirect effect upon physical health,' bemoaned one 1913 report.[11]

In cities, motorized transport was growing apace and, in London, the internal combustion engine replaced the horse as the motive power for omnibuses at a surprising rate (see Table 1).

TABLE 1. Changing means of bus transport, 5-year average, 1900–1930

Year	Horse-drawn buses	Motor buses
1900	3,681	4
1905	3,484	241
1910	1,103	1,200
1915	36	2,761
1920	0	3,365
1925	0	5,478
1930	0	5,953

Source: Philip Bagwell and Peter Lyth, *Transport in Britain 1750–2000* (London and New York: Hambledon Continuum, 2002), 117.

It is not surprising that, in the face of this multi-pronged onslaught from continuing urban growth, the rise of new technology, and the indifference of people who were in a position to do anything about it, people suffered—and, as usual, the newness of some of the noise sources exacerbated the effects of the increased volume. Robert Koch, famous for helping prove the germ theory of disease, predicted that 'the day will come when mankind will have to fight noise just as vehemently as cholera and pestilence'[12]—but clearly that day was still some way off.

So, what did people do in response? As many had done before, those who could afford it retreated—either to the quiet havens of the countryside, or, in the case of Marcel Proust, like Thomas Carlyle before him, to the specially made quiet spaces of the indoors. In Paris, Proust had his room silenced as much as possible by lining it with cork, and spent much of his time in it in bed with the windows firmly closed.[13] Visitors were welcomed only in the relative quietness of the night.

A rather more practical and helpful response came from America in the form of a campaign organized by Julia Barnett Rice, a medical doctor. Initially the target of her ire was not noise in general but one specific example—the so-called social (unnecessary, in other words) use of steam whistles by tugboats in New York harbour. Rice noted nearly 3,000 such sounds in a single night. Despite determined efforts on the part of the tugboat captains to stop her, she collected thousands of signatures of support over the period 1904–5, and finally succeeded in having the Bennet Act passed, restricting such noises. While Proust's response benefited only himself, Rice's efforts led to quieter conditions not just in her own backyard but in harbours across America. And Rice was perhaps the first ever anti-noise campaigner to extend her fight beyond her own personal concerns: encouraged by the success of her campaign, in 1906 she formed the New York City Society for the Suppression of Unnecessary Noise. It was not a marked success in terms of numbers—at its peak in 1907 it had only about 200 members, though many of those were influential businessmen and clergymen—it achieved many notable successes, including controls on the use of 4 July fireworks and the creation of zones of quiet near schools and hospitals.

Rice and her society were by far the most successful anti-noise movement anywhere in the world until the 1960s, and were in many ways ahead of their time, with their careful marshalling of the support of people with power, their concentration on issues that affected large groups of people, and their canny use of marketing, including what would one day be called stunt-casting: Mark Twain—himself an ex-river-boat captain who had no doubt annoyed a fair few people with steam whistles in his time—acted as an honorary head of the children's branch of the Society. It was school visits that drew Rice's attention to the importance of quiet for education, a theme that has continued very strongly to this day.

Elsewhere, other societies were soon set up: in Europe, the London Street Noise Abatement Committee was formed in 1908 and the German Association for Protection from Noise was inaugurated the following year.

The London Committee seems to have had little impact, but the German Association was fairly successful—it limited the use of train signals and factory whistles, and succeeded in introducing quieter pavement surfaces in certain areas. But it did not achieve the national legal changes attained by Mrs Rice's society. Like the nineteenth-century societies, it was composed mainly of academics and artists, and the similarity of its agenda to theirs is clear from a 1908 essay written by its founder, German philosophy professor Theodor Lessing. In it, in terms very reminiscent of Babbage, he describes noise as an act of revenge of the manual labourer against the 'brainworker' who was in authority over him. The title of its periodical underlines the point: *Anti-Rowdy (the Right to Silence): A Monthly Publication on the Abatement of Noise, Rudeness and Uncivilized Behaviour in German Economic Life, Business and Travelling*.

With the fault of noise—according to Europeans—laid squarely at the door of the manual worker, the solution to the problem was obvious: education. In 1900, anthropologist Michael Haberlandt proposed that an eleventh commandment should be taught in every school: 'Thou shalt not make noise.'[14] The view that the working classes like nothing better than making noise meant that noise problems in factories and other workplaces were entirely ignored by the European Societies. And, as well as supposedly being no problem for the workers, machinery noise seemed inevitable, an intrinsic part of modern technology.

So it was people, not machines, that were to be controlled. After all, the machines were the bread of the workers; a young factory worker, when asked at the beginning of the century 'Isn't the noise of the machines awful?' replied, 'Yes, not so much when they are going on as when they stop.'[15]

Meanwhile, however, more and more members of the medical fraternity were waking up to the fact that noise *did* affect workers, even if no one else cared about the fact. In 1907, a quarter of a century after Barr's report, a UK government committee, The Departmental Committee on Compensation for Industrial Diseases, decided after considerable rumination, that—as they might perhaps have gathered from its name—'boilermakers' deafness is unquestionably an injury due to employment'. In 1910 *The Annual Report, Factories and Workshops* added helpfully:

> It is generally known that men employed in certain trades are liable to have their sense of hearing seriously impaired, if not entirely destroyed in the course of time, as a result of long continued exposure to loud noises. One well known instance is that of boilermakers [*sic*] deafness, other occupations are the hammering of metal sheets and cylinders, use of pneumatic tools, beetling of cloth, engine driving and firing of guns.

'Generally known' though these effects might have been, they might as well not have been, for all the action that was taken about their causes: precisely nothing was done about the problem of industrial noise, and nor would it be, for many decades. Even as late as 1935, N. W. McLachlan's book *Noise* devotes a scant half-page to the industrial variety, despite its full title continuing: *A Comprehensive Survey from Every Point of View.*[16]

Not that this neglect was confined to noise: the deaths of workers that were due to industrial injuries were neither properly reported nor collated, nor, in general, was any compensation or change of procedure sought through litigation. Claims for damages were few and rarely successful in the face of a widespread assumption that the life of the worker was cheap and the individuals replaceable—an assumption that was arguably more entrenched in Europe than in the USA. It may well be that Rice's American campaigns were wider in their scope and more successful because of the less class-bound and structured nature of society there.

Street noise was another matter in terms of public concern, but still, very little was actually being done about it, other than the gradual and local replacement of surfaces with quieter materials. In wealthier neighbourhoods with cobbles, the streets were sometimes watered to control noise—which must have been spectacularly unsuccessful and probably continued only because airborne dust was reduced in the process.

Meanwhile, the first ever interest in the application of underwater sound bore fruit in 1901 when the combination of a bell, sounded by compressed air, and a hydrophone—actually just a microphone in a box—was used successfully for signalling between ships. The idea had been proposed as long ago as 1889 but could not be tried out until an effective method of waterproofing the microphone had been developed, by Elisha Gray (who many claim to be the true inventor of the telephone). It was used primarily to warn ships about icebergs and other hazards, and it proved quite successful until it was supplanted by radio direction finders, which were invented in 1912. However, even then, many of these instruments combined an acoustic signal, sent through either air or water, together with the radio signal. The benefit of the acoustic element was that the distances of such beacons could be determined by timing the delay between the radio signal and the sound—just like counting the seconds between lighting and thunder to establish the distance of a storm.[17] Today, automatic underwater acoustic beacons called pingers are used to mark the position of submerged wrecks and dangerous objects. Conversely, there are systems that react to acoustic signals to release underwater instruments, and mines that explode when the sounds of engines approach.

Despite the pioneering efforts of Rice, city noise continued to grow rapidly in the USA, because of both the spread of the cities themselves and the increase in the number of motor vehicles in them. The widely held belief that 'brain workers' were the only ones really bothered by this is belied by the responses of a group of poor tenement dwellers in Philadelphia. When asked in 1915 what the greatest problem they faced was, one woman said:

> 'I speak for every woman here. What we cannot stand is noise. It never stops. It is killing us. We work all day and need sleep and rest at night. No one can sleep till midnight and all the noise begins again at five. Many of us have husbands who work at night and must get sleep during the day, but they get no sleep with all the noise that goes on about us.'[18]

The use of motorcycles, often with no mufflers, was frequently mentioned as a major factor, as was street-car noise—especially when exacerbated by loose junction points. New York was now firmly established as the world's noisiest city, with problems cited at the time being elevated railways, cobblestones, and a new phenomenon—radio advertisements (see Fig. 22). New York has retained this dubious distinction to this day, followed by Tokyo, Nagasaki, and Buenos Aires.[19]

FIGURE 22. New York street cars, 1910.
Museum of the City of New York/Getty Images.

In response, city ordinances were spreading through the country, with Chicago, Cleveland, Louisville, Milwaukee, and New York all establishing quiet zones, often near hospitals, in part thanks to Rice's efforts. However, defining zones was one thing, enforcing them quite another: in the whole of the USA, there was only one anti-noise officer, in the Baltimore police force.

In the UK and in Europe in general, cities were often regarded by transatlantic commentators and travellers as quieter, in part because of the European preference for underground, rather than elevated, city trains. Motorcycles were also much rarer. But that is not to deny that traffic noise was a problem. In many European cities people were becoming very concerned about it by the 1910s. And what really irritated them was the motor horn. As Dan McKenzie, a surgeon, put it in his book *The City of Din* (subtitled with great accuracy *A Tirade against Noise*): 'The motor-horn! The motor-horn! I often wonder why in all the world such an instrument of torture has ever been permitted to exist even for a single day!' and so on at some considerable length.[20] Meanwhile in Switzerland in 1913 a new by-law was passed 'Against unnecessary motor vehicle noise and blowing horns at night'.

Why the bile? Partly perhaps because motor horns were under the direct and immediate control of individuals, and there was a widespread feeling that those individuals were making a lot more use of them than was strictly necessary. The newness of the noise was no doubt a factor too—a hooter in traffic today might be a cause of more amusement than annoyance (and might even be able to do its job without annoying anyone; which would make it an ideal acoustic warning system—at least, until the novelty wore off). Whatever the reason, complaints in the House of Commons became so vociferous that the Motor Traffic (Street Noises) Bill was rapidly steered through both houses in 1911.

Theatres of noise

Meanwhile in Paris, noise was even invading the concert hall—or so the audience of the Théâtre des Champs-Élysées thought when, on 29 May 1913, the first public performance of Stravinsky's *Rite of Spring* took place. It quite literally caused a riot, this being the first time that a really dissonant piece of music had been heard by the general public.

Anyone listening to Stravinsky's music today is unlikely to feel particularly riotous, or indeed annoyed by it. Since perceptions of what music 'should' sound like have moved on so far, it has become impossible to capture the reaction of listeners a century ago, to whom all but a sprinkling of mild dissonance in music could be absolutely intolerable. This is not to say that the definition of dissonance and consonance has changed: dissonance is still often used to express conflict or disturbance in music, while consonances are relatively restful. Usually the progress of the music is from the tension of a dissonance to the stability of consonance. But surely no one today would refer to *The Rite of Spring's* 'almost cruel dissonance', as they did in 1913.[21]

In the same year that *The Rite of Spring* was annoying people in Paris, in Italy a manifesto that politicized noise appeared: *L'arte dei rumori (The Art of Noises)*, by the futurist Luigi Russolo.[22] Like Marinetti before him, Russolo argued that people should celebrate noise—in particular that made by motor vehicles. But he was not content merely to suggest that people should linger by roads and railways to drink in the high-tech ambience; he also proposed, in a far more specific way than had Marinetti, how it should be brought into the sphere of public performance. To this end, Russolo invented twenty-seven noise-making instruments, called *intonarumori* ('noise-intoners') (see Fig. 23). Each instrument consisted of a wooden box with a cardboard or metal speaker attached. The performer turned a handle or pressed a button to produce the

FIGURE 23. Luigi Russolo (left) and his *intonarumori*.
Arena Images/Topfoto.

sound, and used a lever to alter the pitch. Inside each *intonarumore* was a notched wheel, made of metal or wood, that struck a metal or gut string. The name of each *intonarumore* described the sound it produced, so there were 'exploders', 'crumplers', 'hissers', and 'scrapers.' A 1917 performance in Milan, using eighteen such instruments, of Russolo's *Gran Concerto Futuristico* generated a similar reaction to that produced by *The Rite of Spring*, in the form of a barrage of rotten fruit and vegetables from the audience, along with an accompaniment of jeering, shouting, and—appropriately enough—hissing.

In the same year Eric Satie completed the first suite of a kind of music that, for all its small impact at the time, was a better preview of the sound of the future than those generated by the futurists: his *Furniture Music* was intended to be played in the background at such events as civil marriages, receptions, and lunches. A second collection followed in 1920. Satie referred to some of the pieces as 'phonic tiling' and, though they use conventional instruments, with names such as 'tapestry in forged iron' and 'industrial sounds', their presaging of later movements is remarkable.

Meanwhile, many people were finding conventional music quite annoying enough when it arrived via the gramophone, now complete with a noisy amplifying horn—or even two. Gramophones soon became the ghetto-blasters of their day, and did not reside safely indoors on the sideboard for long. Some made their way out into the street, beach, or park in the form of small wind-up devices. And wind people up is just what they did. According to some commentators, this was exactly the point of their conquest of the outdoors—to influence others at a distance, rather than simply to entertain their owners. Anyone listening to a music-equipped picnic or even a long debate by nearby cellphone may well feel the same today.

The first recordings of songs and music (along with speeches, poems, and what their producers fondly regarded as amusing conversations) were made in the late 1870s and increased rapidly in number and popularity through the late nineteenth and early twentieth centuries, despite the lack of electronic amplification. In those first decades, it was the sound, alone and unaided, that had to move the needles that were used to cut a song into a wax cylinder (later a disc).

Furthermore, for many years there was no way to copy the cylinders: a repeat performance was required for each one. Eventually a method was found of copying from a master cylinder by means of a pantograph, which was an arrangement of levers and wires that transmitted the sound vibrations from the stylus that read the master to the one that inscribed the copy, but the masters wore out so fast that the number of copies made was small and the quality rapidly degraded from batch to batch.

So, a successful singer was required not just to sing prettily, but really to belt out his or her hit, over and over again. One of the most popular performers to listen to on the newfangled gramophones was Italian tenor Enrico Caruso, proud owner of a very powerful voice indeed.

So powerful was Caruso's voice that, legend has it, he could shatter a wine glass by singing very loudly at its resonant frequency.

If he really did so, he seems to be the only person who ever has.[23] (Ella Fitzgerald did break a glass with her voice, for a *Memorex* advert in the 1970s, but her personal output needed amplifying from maybe 70 dB to 143 dB to do the trick.)

Once they had made all that effort, many singers were not very impressed with the result—and that was not entirely due to the primitive (albeit ground-breaking) technology on offer. As has become all too familiar to people who have listened to recordings of themselves ever since, the result can be humblingly thin and reedy. This is because, when we hear our own voice while we are using it, a lot of the transmission to the cochlea is via bone conduction. Listening to a recorded version, on the other hand, relies entirely upon air conduction, which augments significantly the sounds over 300 Hz. The difference in the mechanisms is easy to detect if you hum and then stick your fingers in your ears—which makes the hum sound louder.

It was the new interest in iceberg detection following the loss of the *Titanic* in 1912 that suggested that echo-location might be a good way to find such objects underwater (where, as anyone in search of a good analogy for things that are more significant than they appear knows, 90 per cent of each lies hidden). In fact the first related patent—an airborne echo-sounder—appeared within a week of the disaster. An underwater version was patented the following month. They were both the brainchildren of Lewis Richardson, a British inventor. Though numerous other patents followed, the first device that actually worked was patented in 1914, on the eve of the First World War, authored by Canadian-born inventor Reginald Fessenden. He successfully trialled his acoustic underwater detector soon after, detecting an iceberg at a distance of about 3 kilometres.

Despite—or because of—all the noise in the early 1910s, it was a decade of hope for those who suffered, with the rise of anti-noise societies and, through them, of noise regulations. But, in August 1914, all that changed.

10

NOISE IN WARTIME

In London, on the last day before the war, Prime Minister Herbert Asquith recorded that the city was full of the roaring of crowds, the accompaniment to the commencement of hostilities since time immemorial. But the celebrations of war were not to continue for long, as the world hurled itself into its first modern conflict, which rapidly bogged itself down in trenches across north-western Europe. The shattering sounds of the artillery could often be heard in London, so the closer effect on the men in the trenches cannot really be imagined.

Noise was a tool of war as well as its product. To this day, walkers along the south coast of England can see great crumbling concrete dishes pointed forlornly out to sea, looking for all the world like petrified radio telescopes. And they are designed for exactly the same reason—to collect radiation and focus it. But in this case the radiation is acoustic, and the function of the dishes is to spot approaching planes, rather than receding galaxies. The first such 'sound mirror' was cut into a chalk cliff face between Sittingbourne and Maidstone in July 1915.

It may seem surprising that it was thought necessary to build a complex system of aircraft detectors scarcely a decade since the Wright brothers had launched themselves into space on their frail structures of wood and canvas, but aircraft were very much in the news in the 1910s, even if there were not many of them actually in the air. And, while the devices seem not to have detected planes at the time, experiments in 2006, carried out by National Physical Laboratory acousticians for the BBC's *Coast* TV programme on sound mirrors at Denge in Kent (see Fig. 24), confirmed that they could have done so.

Sound underwater

The First World War was in part a contest between competing technologies. German U-boats were a new and terrifying menace against which there seemed initially to be little defence. With UK supply lines compromised and the public increasingly concerned by the invisible threat, the Navy turned to scientists to help, setting up the Admiralty Board of Invention and Research for this purpose in 1915. Among the numerous fields of study pursued—including searchlights and wireless communication—underwater acoustics was included, and it is the research that arose from this that marked the true beginning of the science of underwater sound.

The initial approach was to adopt a suggestion made four centuries earlier by Leonardo da Vinci: 'If you cause your ship to stop,

FIGURE 24. The Denge sound mirrors today.

John Briscoe/Alamy.

and place the head of a long tube in the water, and place the other extremity to your ear, you will hear ships at a great distance from you.'[1] Devices like this, known as Broca tubes, were developed in about 1915, fitted with a rubber diaphragm at one end and a stethoscope at the other. Soon, microphones were added. To begin with, while these tubes could pick up loud underwater sounds, including those made by the engines and propellers of submarines, they could give information about neither the direction nor the distance of their sources. The Board was encouraged, however, to hear that shipping could be detected by the tubes at distances of up to about 20 kilometres away in favourable conditions, and it was also discovered that each type of ship could be distinguished by its sound, from the beat of a drifter's reciprocating engines to the more continuous drone of a destroyer.

Experiments in mounting pairs of such underwater microphones (hydrophones) on either side of a ship's hull allowed a very rough idea of bearing to be determined, but it was the advent of the truly directional hydrophone that turned the systems into something really useful. The first such hydrophones were diaphragms with a microphone in a box in the centre. This design meant that a sound was detected much more strongly if it came from immediately in front of or immediately behind the diaphragm—but, unfortunately, the device could not tell which. However, by fastening pairs of diaphragms together, a unidirectional instrument could be built that was able to ascertain the direction of a sound source to within about two degrees. One of several problems with these devices was that the engines of the ship that carried them drowned out all other noise sources, so it was necessary to follow Leonardo's advice and stop the ship in order to use them. And to ascertain the actual position of the sound source it was necessary to employ a pair of ships and triangulate the results. Still, it was found possible to use this clumsy approach in practice, and some U-boats were successfully located and destroyed by depth charges by this means.

That there was any scientific response to underwater threats at all is surprising bearing in mind that shipping at the time was almost entirely bare of electrical technology: only about 20 per cent of merchant ships using British or French ports at the time had any wireless equipment,[2] and navigation was entirely by lead and line, compass and sextant. Nor was there any background of knowledge of underwater sound to draw on—about all that was known was the speed of sound in water.

Meanwhile, in France, Paul Langevin was experimenting with ultrasonic echo-location. In 1880, Pierre Curie and his brother Jacques had discovered that quartz crystals change shape when an electrical voltage is applied to them, and, conversely, they generate a voltage when squeezed. For the Curies this was no more than an interesting discovery, but Langevin used it in combination with a rapidly oscillating charge to make mica crystal sheets vibrate at frequencies up to 100 kHz—a frequency as unheard of as it was unhearable. At such high frequencies, sound waves are rapidly absorbed by air, but very much less by water. By 1916, underwater ultrasonic signals could be picked up at over 2 kilometres (using a conventional microphone and resonant circuit), and reflections from submerged metal objects at 200 metres could be detected. Working on the Seine and at sea between 1915 and 1918, Langevin demonstrated that submarines could be located with ease. He used a piezoelectric transducer vibrating at 38 kHz, a frequency at which sound waves spread out very little with distance, and this meant that Langevin's system was far superior to Fessenden's audio-frequency version. Quartz crystals continued to be used for many years, but today most ultrasonic transducers use synthetic crystals, often lead zirconate titanate (PZT).

Initially, the idea of using ultrasonic echo-location as a defence against submarines was rejected by the Board because the return signal strengths were so low, but the development of a new type of electronic amplifier combined with a new Langevin design based on a quartz sheet held between two steel plates allowed a much improved device to be built.

Suddenly, the answer to the submarine threat seemed to be at hand, and only two problems faced Langevin and the now very enthusiastic Board: where to obtain large quartz crystals and how to cut them accurately to the required shape. The first problem was solved thanks to a rapid tour of geological museums and jewellers' shops on both sides of the Channel and the second was entrusted to Farmer and Brindlay, a famous tombstone-cutting business in Lambeth.

Thanks to its usual love of acronyms, the military soon decided to call this new technology ASDIC. But thanks to their equally usual love of secrecy, no one seems to have wanted to be pinned down to what the acronym actually meant: there are at least six variants, ranging from the administrative sounding *Anti-Submarine Device International Committee* to the bizarre *Anti-Submarine Division-ics* (with the suffix '-ics' being derived from 'physics').[3]

So promising did this new locating technology appear that an international conference was held in Paris, attended by representatives of the USA, the UK, France, and Italy, to discuss it. Wartime secrecy means that few records or documents of this meeting have survived, but it is clear that very similar developments were taking place in the USA. The conference might well have led to a rapid proliferation of ASDIC technology, but for its dates: 19–22 October 1918. A month later the war was over and no one was interested in submarine detection any more. The equipment was returned to its shelves, the scientists to their laboratories, and the tombstone-makers to their graveyards. Another submarine-detection project was halted then too, when several sea lions that had been trained to find submarines were, in the words of the official memorandum, 'allowed to return to their legitimate business'.[4]

A more aggressive use of acoustics underwater was also nipped in the bud: the acoustic non-contact mine. The first experiments with the devices took place in Parkeston Quay in May 1917, followed by a full programme of development that began the same autumn. In April 1918, the design was considered to be effective, and mass construction began. Many thousands of such mines were ready for laying when the war ended.

It was not only research into the use of underwater noise as a weapon or a defensive measure that terminated with the war—the same was true of practically all other military marine research and development. In fact, so little progress was made between the wars that an official of the Navy Department could only say—in an after-dinner speech in the 1940s—that the Navy had not lost any ground in the period: at the end of the First World War the navy's submarines could do 54 knots underwater and in 1941, at the onset of the Second World War, the navy had submarines that could still do 54 knots underwater!

Another invention designed for the First World War that came to nothing after peace was declared—though it might possibly have done so anyway—was the foot-hooter: a car hooter attached to the sole of the user's boot, so that it hooted at every other step. This German invention was designed to teach soldiers to march correctly, but the inventor suggested it could also be used to cheer up people who were being carried on stretchers, since the bearers could each be equipped with a differently pitched hooter and play the (doubly) unfortunate patient simple tunes as he was carried away. This noisy-shoe idea was reimagined in 2011 in the form of some programmable Crocs that make the noises of a range of surfaces: crunchy snow, pebbles, wood flooring. Rather more usefully they also make discouraging buzzes when the wearer is walking in a direction other than the one programmed for his destination. Maybe this version will even catch on.

The 1920s was a decade of change for many aspects of noise, and in particular it marked the beginning of a properly organized approach to its control. In terms of the beast itself, the largest increase by far was in the noise of transportation. With growing prosperity (prior to the stock-market crash of 1929) there was a rapid growth in car ownership: in Britain the number of motorized vehicles continued its rapid rise as car prices fell: from 150,000 in 1910 to 800,000 in 1920 and 2,200,000 in 1930.[5] Whatever had previously been the case, the motor horn really was a problem now: the government decreed that the UK National Physical Laboratory's acousticians should analyse the sounds of motor horns to track down the causes of their 'stridency', with a view to making a less irritating sound—but careful research revealed that the devices were already as pleasant as they could possibly be: the louder the horns were, the more annoying the sound, whatever the actual type of sound they made. So perhaps it is rather a pity that the car voice pipe was not a success. Invented in 1930 by US driver Eugene Baker, the design was simple: a tube connected an opening at the front of the bonnet to a mouthpiece close to the dashboard. To use it, one simply shouted into the mouthpiece. While this would have had the advantages that one could moderate one's tone and select one's words according to the gravity of the situation (not to mention the class of the person addressed), the effort to be heard above the pressure of the wind blowing down the pipe would probably have been better expended by yelling out of the window in the first place.

Another novel use for the human voice was invented the next year, with a plan to stop snoring by amplifying and playing back the sound of the snores to the snorer, hence waking them up. Perhaps not surprisingly, that didn't catch on either.

11

A TIME OF PEACE?

Immediately after the First World War there was a rapid rise in civil air transportation, drawing on the substantial community of trained pilots and the large fleets of planes that had been constructed for military use. As a result, by 1923 nearly 10,000 passengers were flying between London and Paris each year. In the USA, an airmail service started straight after the war, though passenger flights did not begin until 1926. But from then on, US airlines rapidly overtook their European counterparts: in 1928, German Airlines, the leading European airline, carried 100,000 passengers while US airlines carried 60,000. Just one year later, this latter figure had increased to 160,000, compared with growth to 120,000 by German Airlines. Meanwhile, the number of air passengers on British aircraft rose steeply in the 1920s: from 10,393 in 1922 to 29,312 in 1929.[1] Some of the government were concerned about the noise impact of such growth—not that it might affect people's health or happiness, but that they might complain about it. For this reason, Winston Churchill made it illegal to sue for the noise effects of aircraft in flight in 1920.[2]

From the early days of airlines, the public reaction to aircraft noise was one of intense dislike: people were significantly more bothered by the noise than they were by that from road traffic. Today, a quantity called the Aircraft Malus is sometimes used to quantify this dislike, which makes aircraft noise about as annoying as road traffic noise, which is 5 dB louder—or, in fact, 5 dBA louder (see below for an explanation of this modification). National Physical Laboratory (NPL) tests of annoyance in the 1960s bore this observation out under experimental conditions, finding that a bigger number of interviewees felt annoyed or highly annoyed by aircraft noise compared with road traffic noise at a similar level. It has been suggested that an important component of the Malus is the dislike that people have of noises from above them, as such noises seem both unnatural and dangerous. However, experiments conducted in 2005 to test this had the reverse result to that expected—aircraft noise actually seems to be *less* annoying when heard from above than from in front.[3] So, the Malus remains mysterious. More recently (2008) it has been shown that helicopter noise is even more annoying than that of other aircraft[4]—though, like any other noises, the level of annoyance is very strongly related to the perceived reason for the flight. A celebrity flying overhead is much more annoying than an air ambulance.

Defining the decibel

According to science historians Lorraine Daston and Peter Galison, the late-nineteenth-century downplaying of the significance of human estimates and judgements in favour of measurements—what they refer to as 'mechanical objectivity'—had reached its peak by the 1920s.[5] Such a shift is very relevant to noise, with its endlessly uneasy interplay between the subjective and the objective. And indeed it was in the 1920s that scientists really got their teeth into the quantitative effects of noise, and developed effective ways of measuring it at last.

To begin with, in 1923, the discoveries of the logarithmic nature of the response of living things to stimuli by Fechner, Weber, and Wien were dusted off and tested by Vern Oliver Knudsen at Chicago University. Knudsen investigated anew the mathematics of human response to sound and plotted graphs of what he called 'loudness of the tone'. This quantity was defined as the logarithm of the ratio of the new sound level over the old one, and so was almost identical to what is now called the decibel.[6]

Meanwhile, engineers were grappling with the problem of interference in telephone lines and in particular seeking useful ways to describe the strength of the telephone connection. From about 1904, the *de facto* standard had been that used in the USA, where the science of telephony was best developed. It was the 'mile of standard cable'. To quantify how much the level of a signal changed when some alteration was made to a telephone system, such as the installation of a new telephone, the change could be compared with the loss caused by the addition of a mile of cable. But, since the actual losses were highly frequency dependent, this approach could not usefully quantify the effects of changes that did not match standard cables in terms of their frequency characteristics. So, in 1924, a new 'sensation unit' was defined, which was based on the ratio of the power transmitted through the line before and after the change. The concept of noise as unwanted sound (or, more precisely, 'a sound which is not desired by the recipient') was only at this point introduced into acoustics, by these telephone engineers. While the everyday use of the word had meant 'unwanted sound' since the Middle Ages, acousticians had hitherto been defining it as 'a sound that is not a note', which, apart from including all sorts of quite nice sounds, is not a very helpful definition for those wanting to get rid of annoying ones. This change was a gradual one though: as late as 1961, E. J. Richards, Professor of Aeronautical Engineering at Southampton University, was defining noise as 'everything [the ear] hears that hinders communication'.[7]

In the first attempt to relate audibility to an easily measurable quantity, Harvey Fletcher of AT&T generated sounds in telephone earpieces simply by applying AC voltages to them. Enticing twenty-three of his colleagues into his lab, he varied the level of the voltage until the subject could just hear the sound in the earpiece. Then, he averaged the voltages and obtained a mean voltage equivalent to a just audible sound. Using this figure, he constructed a microphone-based noise-measuring device, with this voltage level as the setting for the threshold of audibility. It was the first sound intensity meter, although not a very good one, owing to a number of shortcomings in Fletcher's approach, in particular the lack of consideration of the effects of different frequencies on audibility level.

In the mid-1920s, a number of international meetings, together with extensive articles and long letters to the *Electrician*, debated the precise definition of such evolving noise units. It was clear that the logarithm of the ratio of the sound of interest to some standard sound was key, but the details were mulled over by many—as was the name of the unit. In the end Colonel Sir Thomas Fortune Purves, Chief Engineer of the British Post Office, proposed a name for the unit defined by the logarithm (to base 10) of the ratio between two sound powers.

Purves noted that the term 'bell' had been suggested as a name for the unit partly for its association with telephone bells and partly because of the popularity of the Greek characters β, ϵ, and l in transmission theory. He suggested that, to avoid confusion with such a commonly used word, the spelling 'bel' be adopted. And so it was. The use of a logarithmic scale was useful not only in that it correctly reflected the way that people perceive changes in level of sound, but also because it made it easy to handle the vast range of sound energies that people are capable of hearing, a noise that causes pain being about a million million times more intense than the quietest audible sound. By 1930 the decibel scale was pretty much universal, leading the UK's *Saturday Evening Post* to say that 'the fight against wasteful racket is out of the hands of cranks and theorists and is being directed by trained technical minds'[8]—a view that probably pleased the telephone engineers more than the scientists at the time.

Unfortunately, though the use of the decibel has the advantages of compactness of expression of a very wide range of sound levels, and relevance to human hearing, it is really rather fiddly to use. In particular, while units like kilograms can be added together simply by summing their values, the same is not true for decibels: if two similar 10 dB sounds are heard together, the result is not a 20 dB sound, but a 13 dB one. Also, as decibels refer to ratios, they are meaningless unless the denominator of the ratio is defined: 20 dB means 'a hundred-fold greater power'—but what is the quantity that is folded? What, in other words, is 0 dB? Usually it is the threshold of hearing: the level that can just be heard by a listener with good hearing. But this is not always the case, which, as will soon emerge, can lead to great difficulties.

A complication is that the ear does not respond equally to all frequencies: it is about 1,000 times as sensitive to a 1,000 Hz tone than to a 100 Hz one. Hence, devices that measure noise levels must be furnished with filters (electronic or software-based) that mimic the frequency response of the human ear. A number of weightings are in use, but the most popular since the 1930s is A-weighting—A-weighted decibels usually being written as dBA. Over the next few decades dozens of differently weighted versions were defined—even including special weightings for different types of dog. In the 1970s the international acoustical community decided that this was getting a bit out of hand. Many of the weightings were then discouraged, with only the A-weighting being fully supported. (C-weighting also remained fairly popular: it does not reduce the low-frequency component as much as A-weighting does and so can be more relevant to the impact of sounds like aircraft noise that have a significant low-frequency component. While the ear is relatively insensitive to this component, it is the cause of structural vibration when aircraft fly over a dwelling, and it is this vibration that causes a large part of the disturbance to residents.) Nowadays decisions about what type of weighting to apply are of less importance, since cheapness and compactness of computer memory mean that data can be stored as actual pressures, and these data can later be converted to logged values with any weighting one could possibly wish for. In earlier decades, it was only the weighted data themselves that could realistically be collected in the first place.

In particular fields, special weightings are still vital—for instance, marine animals have radically different responses to different frequencies than we do, so to use an A-weighting for underwater sound when one is studying its effect on, say, whales, would grossly distort the results, and, consequently, special whale-based weightings are in regular use.

There is a final complicating factor. A hearer's perception of noise depends on his or her distance from the source—an obvious enough point, but the corollary is that the power of a noise source is meaningless as a guide to how loud it sounds unless the distance to that source is stated.

In view of such complexities, it is just as well that 1928 saw the organization, at long last, of the many scientists and engineers who were working in the diverse range of topics within acoustics. As the scope of science and technology became broader, the number of American Physical Society members who were interested in acoustic papers became proportionately smaller, until Harold Fletcher—who had been pushing for the use of what would later be called the decibel scale since 1923—became so disenchanted with the lack of interest caused by one of his presentations to the Society that he set up The Acoustical Society of America (ASA), the first of its kind in the world. Its first official meeting was held in 1929.

The formation of the ASA was timely, as the late 1920s saw not only a lot of arguing about units but also an acceleration of research in several areas of acoustics, from noise reduction to ultrasound—then sometimes called supersound (or, very confusingly indeed, 'supersonics'). One of the most important characteristics of ultrasound, which is not generally shared by the hearable variety, is that it can interact with matter, providing information about it at low energies or changing it at high ones. This discovery was made by G. W. Pierce in 1924, when he noted that 'supersound' was modified—dispersed—when it passed through carbon dioxide. If there could be interactions between sound and gases, what about solids and liquids? What about people? In 1926 and 1927 a range of experiments of the effects on samples of living tissue was carried out at Tuxedo Park, New Jersey, and showed just how easily such samples could be destroyed. Fish and small animals met similar fates. Further experiments revealed that ultrasound had thermal, chemical, and photochemical effects and could also produce stable emulsions between immiscible fluids such as mercury and water, or milk and cream (hence allowing the production of homogenized milk). It could destroy viruses and bacteria too. The physical nature of ultrasound is different, too: while it had previously been regarded merely as unhearable sound, it was now realized that these unique properties of ultrasound could make it very useful indeed, especially since it is highly directional, and so can be aimed accurately. While it travels only a short distance in air before being absorbed, ultrasound can be transmitted over quite long distances underwater. And its inaudibility means that it can be used at high powers with no need for special sound insulation or hearing protection.

In the audio range too there was a flurry of acoustical research in the late 1920s and 1930s. A great many experiments were tried out in order to quantify its effects, thanks to the sudden availability of new instruments made possible by the rapid development of electronics. So, after thousands of years of slow progress, the science of sound leapt forward. The many studies that were undertaken unearthed some fascinating—though maybe not very useful (or indeed, repeatable)—results. For instance, subjects were asked to sit cross-legged and listen to different types of music, while their knees were hit with little hammers. The amplitudes of the swings of their lower legs were then carefully measured. While a quiet room elicited a swing of 7.4 centimetres, Beethoven's *Funeral March* resulted in 10.5 centimetres but Chopin's *Raindrop Prelude* generated only 6.8 centimetres. What did this mean? No one seemed to know. Another vital experiment revealed that listening to jazz increased the strength of the grip, and yet another compared the effects on animals before and after their brains were removed. Human subjects were fed balloons that were inflated to fill their stomachs, and changes in patterns of variation of the balloon pressure were then studied as the person was exposed to loud noises. Oddly enough, this idea caught on, and several acoustic professors encouraged their students to swallow such balloons and then frightened them in various ways. Perhaps not surprisingly, they 'found the stomach rhythms of students profoundly affected by an unexpected pistol shot or by touching a snake in a dark room'.[9]

Rather more usefully, typists had their breath analysed while typing in quiet and noisy environments and their energy expenditure was calculated from this. A 'noise machine' was switched on some of the time and its effect on their energy expenditure determined. It was found that it required an extra 38 per cent to type when the environment was noisy. The subjects had to wear special masks while working—and while sleeping too, so as to get used to them (see Fig. 25).

It is only fair to say that in the 1970s, and again in 2006, some studies actually found the converse—that typing speed can be *higher*

FIGURE 25. All in the cause of science: a typist, now completely used to wearing a mask thanks to being forced to sleep in it for nine months, is carefully monitored by a researcher.

National Institute of Industrial Psychology.

in noisier conditions, though the 2006 study also made it clear that noise increases fatigue, so is hardly to be recommended.

Lion in the city

By the 1920s, the noise problem in cities was at such a level and had received such publicity that it finally attracted serious official attention. There was widespread pressure for a more thoroughgoing set of legal instruments and regulations, to augment the piecemeal and largely inefficient by-laws and rules then in force. Up to this time, all such efforts had been held back by the lack of suitable measurement techniques: the only objective measurements of any relevance that had ever been made were those of Barr decades before, and in the 1920s the only method that was available was the use of the newly invented audiometer, which was actually designed to test hearing.

The operating principle of the audiometer is that of sound masking: to measure the level of a noise, an electronic buzzer of a similar pitch is listened to through a single earphone while the sound of interest is listened to with the spare ear. The volume of the buzzer is reduced until it is not quite possible to hear it. A more primitive version of the same masking approach used tuning forks. A fork was struck in a standard way and held close to the ear, and the time until the tone became inaudible measured on a stopwatch.

The need to select suitable buzzer frequencies meant that the results obtained by audiometry were time-consuming as well as rough—and they still depended on the judgement of the user. But the development of the condenser microphone from 1915 to 1920 paved the way for modern measuring devices, and work to develop the first sound level meters (then called acoustimeters or noise meters) began in the mid-1920s. These meters were simply amplified microphones connected to a display through a series of filters that approximated the frequency response of the human ear. Though probably the biggest breakthrough in the whole history of noise control, they were not the easiest gadgets in the world to use, partly because they were rather enormous. A van and at least a couple of robust scientists were essential actually to get them anywhere noisy (see Fig. 26).

So, armed at last with sensible units, useable instruments, political will, and even a bit of funding, scientists could at last tackle environmental noise, and the first stage was to find just how much of it there was. So, from November 1929 to May 1930 the first proper environmental (or 'community') noise survey was carried out, in New York City. There were two main aims: to provide a city noise benchmark for future comparisons and to determine the magnitudes of specific sources so that the effects of any improvements that were made to them could be measured. Along with the newfangled noise meters, tuning forks and audiometers were used. The unit was the newly accepted decibel (not yet specifically

FIGURE 26. The three devices shown in front are an objective noise meter, analyser, and heterodyne oscillator. Behind them are the three battery supplies that helped to make them even less portable. Here they are shown measuring the noise made by a car engine. Actually to transport than anywhere, a much bigger vehicle would have been needed.

S. S. A. Watkins/Western Electric Company.

A-weighted, but the frequency weighting networks built into the devices were similar).

Noise levels at ninety-seven outdoor locations were measured by a team of acousticians in a van (which was essential to avoid carrying the audiometer around, since it was not much lighter than a noise meter, and also rather delicate). In addition, a special study was made of that notorious irritant, the motor horn, of which it was found that 42 per cent were louder than they need be to override the loudest average outdoor noise levels, 45 per cent were just right, and 13 per cent were too quiet.

Among the many individual sound sources measured were steamship whistles, elevated trains, church bells, and—for no

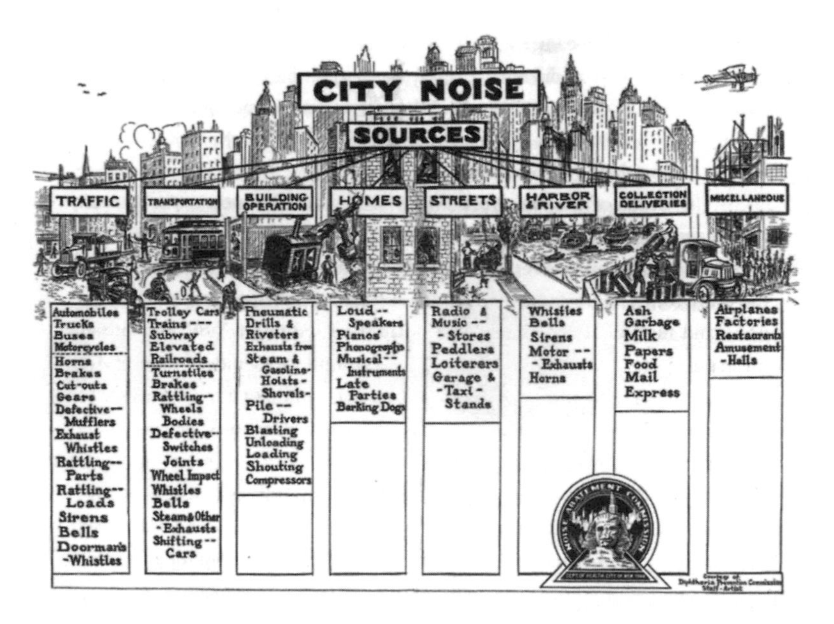

FIGURE 27. The noises of New York City, as collected in the world's first noise survey. The vast number of traffic sources has been accommodated by artificially dividing them into two columns—and still some crop up elsewhere.

New York, Noise Abatement Commission, Department of Health, 1930.

particular reason—a lion. The loudest noise recorded was that made by hammering steel plate (see Fig. 27).

The survey was quickly followed by similar ones made in several other cities, including Chicago, Boston, London, Paris, and Berlin. In New York itself, with evidence to rely on at last, the city set up a Noise Abatement Commission and carried out another survey in 1930.

Noise complaints were also properly collated and analysed. Rather to everyone's surprise, the elevated railways and streetcars of New York were, respectively, only the second and third most significant noise sources. By far the main one, accounting for 10 per cent of noise complaints of all types, was lorries—and so it is to this day. Part of the reason is that, while lorries generate only about 10 times as much mechanical power as do cars, they can often emit 100 times the acoustic power. Diesel engines are louder than petrol ones because the combustion and hence pressure increase is more rapid—and, in addition, such engines make about the same noise, irrespective of their load.

Rude noise

By 1930, vehicles powered by internal combustion engines had become the predominant mode of transportation and source of urban noise in most Western cities and during the rest of the decade the trends of noise increase that had begun in the 1920s continued: in England the number of licensed cars grew from 474,000 in 1924 to 1,944,000 in 1938,[10] and there was little modification to road surfaces or road layouts to accommodate this increase. As a result, noise complaints soared, peaking in 1936, when there were 12,480 complaints[11] (nowadays there are typically 3,000 to 4,000 complaints annually, despite there being about ten times as many licensed vehicles). The German author Robert Musil commented at the time that the cars made 'a wiry noise, with single barbs projecting, sharp edges running along it and submerging again and clear notes splintering off—flying and scattering'.[12]

The rising *level* of noise, however, was not itself criticized greatly at the time. In fact, the chairman of the Anti-Noise League, Lord Thomas Jeeves Horder, commented that people could 'stand the blended sounds of traffic when these make a general hum', but 'if, on top of this, an unsilenced car or motor-cycle accelerates down the street, not only is our sense of hearing, but our sense of justice, is outraged'.[13]

Both in the UK and the USA, the debate about what to do about noise—or, rather, noise-makers (especially hooter-users)—continued, in learned societies and in newspapers alike. The consensus of the vociferous was not that noise was bad per se but that it should not be made as an expression of individual feelings. In both these countries, and in Germany and the Netherlands too, the message was that an orderly society was the answer—good noise control was a matter of politeness and good taste. 'Needless noise is bad manners', said the Anti-Noise League in 1938, and the second New York noise report, published in 1932, concluded: 'We cannot expect quiet until the millions realize they have sold their birthright for a radio and an automobile. Therefore, the only permanent contribution the Noise Abatement Commission can make is to assist in educating the people.'[14] This was exactly the same conclusion that had been drawn a generation earlier. This time, however, there were more attempts to put it into practice. Noise abatement societies in Chicago, the Hague, and other cities organized 'silence weeks', and special exhibitions and many pamphlets were distributed, urging people not to sound their horns unnecessarily (see Fig. 28). There is nothing to suggest that people did try to be quieter as a result. However, the publicity resulting from these activities led to laws against unnecessary hooting of horns and/or design limitations on their performance being passed in France, Italy, the Netherlands, Hungary, Romania, Germany, the USA, and the UK.[15]

On the other hand, to a certain extent the growing noise of cities was welcomed as a sign of their growth in the face of the Depression: 'Isn't it precisely . . . the big noise, the detonation of our

(a)

FIGURE 28(a). Anti-Noise League poster from the 1930s.

Science and Society Picture Library/SuperStock.

national dynamite, that attracts the big crowds which make New York?' asked a New York newspaper in 1932.[16]

Meanwhile indoor noise increased markedly in the 1930s both at home and at work, thanks to the continued rise in popularity of the

(b)

FIGURE 28(b). Anti-Noise League periodical from the 1930s.

Anti-Noise League

radio and the gramophone. In the UK, this was greatly assisted by the introduction of a fully functional National Grid system in 1938. Radio schedules became fuller and fuller until, by the 1940s, there was continuous transmission on some channels.

Although the number of aircraft in use in the 1930s was tiny compared to today's figures, noise was increasingly perceived as a problem there too—and not just to those the craft flew over. Though still available only to the wealthy, air travel was now being considered as a plausible alternative to long-distance journeys by road, rail, or ship—but the noise issue put potential passengers off. A 1934 report explained:

> when air lines in this country started... carrying many passengers, who flew as a quick method of transportation rather than merely for a thrill, then the passengers begin to complain... There are now many people who are air-minded, but the majority cannot be converted to use air transportation until it is brought out of the discomfort level.[17]

And the air-minded were quite right to be concerned. The same report included measurements made in many popular aircraft, including the Ford tri-motor, which turned out to be

> an example of what complete neglect of acoustic considerations may produce... Prolonged exposure to this high level of sound can temporarily injure the sense of hearing. Ringing of the ears and physical discomfort persist for several hours after being subjected to this noise level. A long period of subjection to this level causes permanent hardness of hearing.[18]

While the public and the lawmakers were becoming increasingly concerned about noise levels, the reaction from the medical community was considerably more muted. There were isolated presentations by doctors and others concerned about the effects of noise, but in some cases these seem designed more to alarm than to inform: even the humble (and in those days omnipresent) brown paper bag became a menace according to neurologist Foster Kennedy, who, in a 1930 radio broadcast, claimed that sudden noises like that of a 'blown-up paper bag... raise the brain pressure to four times normal for seven seconds and keep it above normal for 30 seconds... If this is what an innocent paper bag explosion does to your brain, what does an unmuffled motor truck do to you?'[19]

What indeed? Not a lot, according to psychologist Frederic Charles Bartlett, who, in 1934, claimed:

> Where objection is to noises we can suspect, with good reason, a psychological and special history of the trouble. This is the case again and again, how often only a statistical survey which has not been carried out could reveal. A man tired, run down, bored, maladjusted, uninterested, seizes upon anything outstanding from his environment to explain to himself and to others the unsuccess that life has brought him. Noise, as we have seen, is just one of the things which, by its qualities, stands out prominently on almost any background. So it is not too much to say that whenever, in any community, a sweeping and passionate condemnation of noise is popular, there, within that community, are almost certainly a lot of people who are ill-adjusted, worried, attempting too much, or too little. The complaint against noise is a sign, sometimes, of a deeper social distress.[20]

This dismissive view was echoed by N. W. MacLachlan, a doctor of engineering specializing in noise, in 1935: 'In general, when a breakdown in health occurs under exposure to noise, there are other influences at work. The noise seems to act as a catalytic agent or accessory factor, thereby inducing or accentuating a nervous state.'[21]

The concentration by campaigners and the focus of lawmakers on specific sources of noise, on the one hand, and on general recommendations to be more considerate, on the other, may seem rather strange to us today, given that neither was likely to be very effective in stemming the rising tide of noise in cities, but the issue of a high noise level can be tackled only when that level can be measured. As Lord Kelvin (William Thomson) said: 'when you can measure what you are speaking about, and express it in numbers, you know something about it; but when you cannot measure it, when you cannot express it in numbers, your knowledge is of a meagre and unsatisfactory kind.'[22] Without adequate measurement, one could neither quantify the problem, nor check the success of attempts to reduce it. Though this measurement process had begun with the New York noise survey in 1929 and the others that followed, it took some time for their influence to spread, partly because of the unreliability, expense, and general cumbersomeness of the equipment, but partly because it was realized that the decibel, A-weighted or plain, was not really an adequate unit for the task. What was needed was a way to quantify not just the physical level of noise, but its impact on people. While high noise levels can cause physical effects, by far the largest consequences in terms of numbers affected are annoyance and sleep disturbance. The relations between these effects and the sounds that cause them are not only complex; they are also very personal. A piece of jazz that sends one person peacefully to sleep may have the opposite effect on another. The scientists of the 1920s and early 1930s, in pinning their hopes on the decibel, had not fully realized what a can of noisy worms they were opening—hence, perhaps, the emphasis on such things as the dire effects of popping paper bags on 'brain pressure', which at least were nicely measureable, even if of dubious relevance. Clearly, a great deal of further work was needed to make quantitative such effects as annoyance and relate them to directly measurable parameters, and so this work proceeded apace. Was there, scientists asked, *any* measurable parameter that correlated with annoyance? Though several candidates were suggested, by far the best-accepted and quantified was loudness. Once this was well established, it was still necessary to express the perceived loudness of sound in some unit that was simple enough to be widely used while accurate enough to capture the problem.

Hence, in the 1930s, several new units were developed, including the sone and the phon, both of which attempted to quantify perceived loudness. As it was now well appreciated that pitch is no more equivalent to frequency than loudness is to pressure level, the mel and the bark were also defined, to capture the subjective pitches of sounds.[23]

These units *should* be very useful ones: sones, for example do accord more with the impression of loudness than do decibels; a supermarket is about 7 sones, and a busy restaurant or canteen about 15 sones—and one might perhaps agree that the latter is around twice as loud as the former. The interior of a suburban train is around 30, and the doubling of perceived noisiness seems about right here too. Conversely, the A-weighted decibel levels of these three environments are around 55, 65, and 75.

For a time from 1954 onwards, the sone became the noise unit of choice for many, but this turned out to be only a brief flirtation. While it is still occasionally referred to (for instance, sones are often used to describe the effectiveness of 'acoustic' carpet underlays, but, as an explanation of the unit is rarely included, one wonders how informative this really is), most measurements of noise that are relevant to human hearing are made in decibels—usually A-weighted ones (though the 'A' is frequently lost between measurement and media, and the units are frequently not understood, and hence misused, by non-experts). The watt has its adherents, especially in the world of underwater acoustics, and decibels are rarely used to describe the powerful sounds used in ultrasonics either.

The fact that the decibel is frequently used—and then often misquoted—in contexts where it has serious shortcomings means that the fight against environmental noise must be carried out with a blind spot—a permanent one, judging by all the progress that is being made on either the understanding of the decibel or the use of more sophisticated measures. Which is none at all.[24]

Routes to silence

Despite the practical limitations of noise measurement in the 1930s, its law-enforcement implications were seized on with great enthusiasm by Bakker, the Head of the Amsterdam Police. Bakker had been approached by Zwicker, an acoustical engineer who had developed a simple noise-measuring device called the *Silenta*.

So impressed was Bakker by Zwicker and his device that he launched the Dutch Silence Brigade, with the intention of using the *Silenta* much like a breathalyser—to identify scientifically not only who were the loudest motorists but just how noisy they were.

The Brigade took to the streets of Amsterdam in 1937 and comprised four police constables together with a police inspector, equipped with a motorbike and side car, with which to pursue (as quietly as humanly possible) errant noisy motorists.

Unfortunately, the *Silenta* was not anything like as simple to use as Zwicker had hoped: to obtain a satisfactory reading, it had to be positioned at least 7 metres from the noise source, in a flat and open space. Since Amsterdam's streets and bridges are picturesque, verging on the cramped, such spaces are in short supply, with the result that miscreants had to be accompanied out of the city to open areas where the measurements could be made—if it wasn't windy there, as it usually was.

Nevertheless, Bakker did not lose faith in the machine, and of course the knowledge of its existence and that of the Brigade may well have been enough to discourage many noisy drivers. He even planned to extend it to the area where enforcement had long feared to tread: neighbourhood noise. But the customary reluctance was well founded. To assess the noisiness of someone's gramophone or radio, it would be necessary to enter the home, and the public and press baulked at this: for one thing, it was widely held that people had a right to do what they liked in their own homes, and it was also very clear that reactions to many such indoor noises were highly subjective. Mocking accounts of imagined encounters appeared in the Dutch press and the extension of noise policing to the home never happened.

Meanwhile, there was the first occurrence of an often-repeated attempt to deal with noise in itself, with the patenting in 1933 of Paul Leng's idea of anti-sound. The principle is simple: sounds are waves, and, as with any other type of wave, when two identical waves travelling in opposite directions meet, they will cancel out when the troughs of one meet the peaks of another. So, two sounds can produce silence, and in principle we can fight sound with sound. No practical use of this idea was made by Leng himself, but it was only a few years before the first experimental systems were made. And they

FIGURE 29. Wave interference.
Sybille Yates/Shutterstock.

have continued to be made ever since. Unfortunately, despite many years of research, the laws of wave interaction themselves mean that the possibilities of sound cancellation (or active noise control, as it is often known today) are rather limited. For a start, one needs to have a very good knowledge of the noise source if one is to mimic it, so in practice the technique is limited to continuous noises that vary little, or slowly, in terms of frequency content and level. But a more fundamental problem is revealed simply by watching the interference patterns of, say, an object dropped into a still bath of water. With the right lighting one can see the interference pattern generated by the waves and their reflected versions on the bottom of the bath (see Fig. 29). There are indeed some areas of flatness where the waves cancel out—but there are also higher points where peaks meet and add to other peaks, and so increase the water level. In other words, the interaction of two waves of the same length and strength is usually a pattern of both destructive and constructive interference—acoustically speaking, these would be zones of relative quiet and of increased noise. While it is possible to set up loudspeakers so that sound is exactly cancelled, this requires very special geometries and precise frequency-matching. The technique can nevertheless be useful, even when a pattern of louder areas is formed, if the system can be arranged so that these areas form in positions where there is no ear to hear them. So, one can indeed reduce the sound level at the position of a pilot's head in a plane cockpit (such systems work best when integrated into headphones and are also very effective in helicopters with their low-frequency blade-passage noise), but it is not the route to quieten houses or cities.

At the same time, practical ultrasonics received a boost from technological developments: in earlier experiments, ultrasound had been generated by supplying a flat piece of quartz with a rapidly varying electrical current, to which the quartz vibrated in response. But in 1935 the first concave vibrating quartz plate was constructed and used successfully to focus ultrasound beams. This was doubly valuable, in that it could not only increase the intensity of the sound but also be used to focus it at a particular distance from the plate.[25] It was soon realized that such focused beams offered an alternative to surgery, as the ultrasound travelled easily through living tissue. Experiments on blocks of paraffin and pieces of liver revealed that it was indeed possible to heat the inside of the material without greatly affecting its surface, and the experiments were extended to animals. The primitive equipment used was not very effective, and the poor dogs and cats who were used in the experiments suffered unintentional skin burns as well as deliberate brain damage—though in all cases the experiments were at least stopped quickly enough for the animals to recover. The technology is now routinely used to shatter kidney stones (a technique called lithotripsy), and can also be applied to treat otherwise inoperable brain cancers—for this, an ultrasonic helmet is worn, and ultrasound enters the brain through an entire hemisphere. In these High Intensity Focused Ultrasound (HIFU) applications, tumour cells can be raised to temperatures of over 70°C, while those surrounding them are almost unchanged—and, unlike chemotherapy, no dangerous residues are left in the body. Unlike radiotherapy, there is no increased risk of other cancers either. In Korea, the range of applications of focused ultrasound—guided to its target by means of Magnetic Resonance Imaging (MRI)—is being broadened in clinical trials to include such illnesses as prostate cancer, essential tremor, and even obsessive compulsive disorder.[26]

So, despite the partial blind alleys of the *Silenta* and of active noise control, and the lack of interest in workplaces, the 1930s was a time of real progress against noise, with acceptable units, just-about-adequate measuring devices, a great deal of public interest, scientific research, and political activity. However, events in Europe were soon to lead to a wholesale abandonment of anti-noise activities in that continent.

12

A DIFFERENT KIND OF WAR

Once war seemed inevitable, many people decided anti-noise regulation was irrelevant: in 1938, *Punch* published a fictional plan to deaden the sound of bombs, and the British Anti-Noise League itself said its campaign against aircraft noise was 'unsuitable' given the need for rearmament. In 1940, the Dutch Anti-Noise League stopped work too.

None of which is to say that noise research was ended by the outbreak of hostilities. On the contrary, just as in the First World War, noise was a key element. This time, it was to be used to attack as well as to defend.

Hitler himself was clear about the benefits of loudness when it came to speeches, and he is said to have declared in 1938: 'We should not have conquered Germany without the loudspeaker.'[1] The radio too was key to both communication and propaganda on both sides.

A rather different use of noise for warlike purposes was made by the Sturzkampfflugzeug dive-bomber—or, as it was more widely known, the Stuka. The Stuka was a two-man German plane, which first flew in 1935. Though it had been used during the Spanish Civil War in 1936, its fame—or infamy—springs from its role in the London Blitz in 1939–42. The plane's design was highly advanced for its time, incorporating an automatic system to pull it out of its bombing dives even if the pilot lost consciousness from the acceleration. It was also equipped with a 'Jericho Trumpet', a loudly screaming siren intended to intimidate its target populations, just as war cries had done for millennia.

Other weapons of the war were signalled by either their noise or the disturbing lack of it. V-1 flying bombs—or doodlebugs—were simply unmanned planes directed over London. They flew until their fuel ran out, at which point they fell from the skies and detonated. Consequently, their distinctive noise was listened to with rapt attention by everyone in earshot—if it stopped, an explosion was only seconds way. Conversely, the supersonic V-2 rocket weapons deployed towards the end of the war arrived with no audible warning at all.

Developments underwater

Although fundamental research in the development of underwater signalling and sonar devices was halted between the wars, attempts were made to apply the knowledge that had already been gained, through experiments with ultrasonic echo sounding. To begin with, signals were fuzzy and hard to interpret, and it was acousticians in Norway who had the first clear successes in the use of the technique to detect shoals of fish: sprat in 1934 and cod in 1935. It is said that one of the fishermen recruited to test the detector frequently regaled his colleagues at the pub with his poor opinion of the technology/scientific team/whole idea, and how he had no choice but to go on testing it for just a while longer—explaining his record-breaking catches meanwhile as a mere coincidence.[2] Echo-based techniques for fish stock location and measurement have been routinely used ever since. In 1930, a new echo sounder was developed by the Admiralty Research Laboratory (ARL) for the purpose of mapping the seabed, and successfully tested off the coast of Sheerness. This new technique too was rapidly adopted internationally and quickly replaced the arduous and antiquated lead and line system that had been used for centuries.

But it was the renewed outbreak of hostilities that provided a real burst of interest in underwater acoustics. Though ASDIC research had ceased at the end of the First World War, it had not been forgotten, and, at the commencement of the new war, military top brass returned to it. More sensitive hydrophones were developed and became so effective that they could pick up a wide range of ambient underwater sounds. But few of these came from ships. A complex and baffling menagerie of noises was revealed, mostly of unknown origin, and so a programme of underwater noise research was initiated in many countries, but by the USA in particular. A ground-breaking paper published by an American military research team described for the first time the types and—insofar as they had been identified—the sources of underwater sound. Many strange phenomena were uncovered: it was found, for instance, that croakers, which make sounds, 'resembling 4 to 7 rapid blows on a hollow log',[3] take part in a regular dusk chorus along the Atlantic coast of North America, just as some birds take part in a dawn version. One of the most puzzling unidentified noises occurred each evening before midnight and was described, rather unhelpfully for a modern reader, as being like 'the discordant sound of the peanut vendor's whistle'.[4] Whatever it was, it was 15 dB louder than the background noise and concentrated at 3 kHz, and it masked all but the loudest sounds of shipping.

The very existence of a complex underwater soundscape was unsuspected before this time—the 'deeps' were proverbially silent, and Sir William Bragg, an expert in the field of acoustics, had even commented in 1920 that 'the silence of the sea no doubt goes with the fact that fishes do not make much, if any, use of hearing'.[5] As the wartime acousticians quickly discovered, it is actually very much noisier under the water than above it. The fundamental reason that sounds from under the sea do not reach airborne ears is that the surface of the ocean is an acoustic mirror, reflecting back almost all the sound from beneath.

Another strange discovery was the identification of a sound whose pressure was out of all proportion to its source—which turned out to be the snapping shrimp, a small crustacean with an intimidating claw (see Fig. 30). When the claw is snapped, the

FIGURE 30. The snapping shrimp *Synalpheus hemphilli Coutière, 1909*, brandishing its noise-producing claw.

Courtesy of Dr Arthur Anker.

short-duration sound produced is powerful enough to kill any small fish in the immediate vicinity. The shrimp hunt by lurking in burrows on the seabed, with their antennae exposed to detect any motion caused by passing marine life. The shrimp then emerges and snaps its claw, after which it drags the stunned or dead fish back into its hole to feed. The snapping sounds are also apparently used for communication. Maximum peak-to-peak source levels can be as high as 189 dB, and the noise includes components of up to 200 kHz (it peaks at about 2 kHz). It may be that the use of short-duration sounds for stunning prey is not confined to small animals like the shrimp: the sperm whale can generate a 'click' (though the word hardly seems adequate) whose peak-to-peak source level is a massive 243 dB. This is higher than most military sonar systems.

The development of ASDIC, now rebranded as SONAR (Sound Navigation and Ranging), was rapid, and it soon became a highly accurate and effective system. German researchers strove to develop a defence and came up with anti-reflective coatings, now called stealth coatings, in the form of a layer of Alberich tiles. They were made of Oppanol, a new synthetic rubber compound, and they not only muffled the sound of the engine but also hindered active sonar by absorbing the sonar pulses. The tiles were first used operationally in 1940 on the U-67, but there was no suitable adhesive that could withstand both the flow of water and the effects of changing temperature. It took until 1944 to develop adequate adhesives, and the tiles were then used to cover the U-480. The submarine did sink four allied ships in 1944, but whether its prowess was due to all the research and retiling is unclear—the tiles reduced sonar signal reflections by only 15 per cent.

The discoveries that the American research teams made during the Second World War went far beyond immediate military concerns. In particular, it had been speculated for some time that low-frequency sound might propagate over significant distances through the ocean, and in 1944 the idea was tested—with startling results when a mere half-pound of TNT was easily detected by a hydrophone at a distance of over 1,300 kilometres. The effect is due to a worldwide subsurface ocean layer in which sound waves are trapped, because of the pressure and temperature conditions above and below it. In this so-called deep sound channel, where sound-wave velocity is a minimum, sound waves can travel only laterally—so they do not die away until great distances have been traversed. The sound itself is changed markedly as it travels through the channel, so that what starts as a bang soon becomes a distinctive accelerating series of pulses.

This deep sound channel is used for a technique nicely acroynmed SOFAR (Sound Fixing And Ranging), which was effectively launched in 1960 by the Australia–Bermuda Experiment, in which explosions were made off the coast of Australia and detected 20,000 kilometres away in Bermuda. Today, SOFAR is used as an accurate method of measuring global sea temperatures, an important parameter for determination of global warming: in this system, called Acoustical Thermometry of Ocean Climate (ATOC), low-frequency (75 Hz) signals are transmitted over global distances, and their changing arrival times give an accurate measure of overall sea temperatures.

ATOC nearly never happened because of a misunderstanding of units—a 250 W loudspeaker, actually rather quieter than a single tanker (with which the part of the ocean chosen for the tests was thronged) was thought to be so loud it would harm whales and other marine life. The mistake was a simple one—a confusion of units led to the prediction that the speaker would actually radiate 250 *million* watts, which would probably have caused an enormous explosion, could such a thing be built and switched on. But such is the power of the media that a *Los Angeles Times* report[6] set off a hysterical reaction that ended up delaying the project for years, no matter how many times the error was explained.

Grin and bear it

The war affected noise in many other ways. By the time it ended in 1945, the UK was an impoverished country with strict government control and a population trained to respect authority and to put up with things. In consequence of the poverty, there were few cars; in consequence of the prevailing attitudes, few noise complaints. There were, however, an enormous number of trains—in fact more people travelled further in the UK by train in the 1940s than at any other time. It was against this background that, in 1948, the first survey of UK attitudes to noise was carried out.[7] It revealed that, for many, noise was simply not an issue. Questioned about traffic or industrial noise, 58 per cent of those surveyed did not notice such noise at all, while 23 per cent were disturbed by it. The figures for neighbourhood noise or noise generated in their own homes were broadly similar, with 60 per cent not noticing any such noise, and 19 per cent being disturbed by it. These figures are significantly lower than those found by most noise surveys conducted since. In fact, in the most recent (2008) survey 26 per cent of people were disturbed by neighbourhood noise.[8]

When noise *was* complained about in the immediate post-war period, whether in the UK or in the USA, the official response was, to say the least, rather cursory. Thanks to the rather cavalier attitude to noise problems of pilots and their controllers (not having been taught that there were any), the response to local complaints was often limited to putting up signs saying things that today would be thought of as at best useless and at worst inflammatory—like 'Listen to the sound of peace and security' or 'Experience the sound of security'. One even included a picture of an F105 *Thunderchief*, a supersonic fighter-bomber well known—as its name might suggest—for its ear-shattering noise.

Another significant development in the late 1940s was the introduction of portable tape-recorders. For the first time, it became possible actually to capture the salient features of soundscapes. Many of the earliest such recordings—in particular, of London street sounds—were made by Christopher Stone, who had become the world's first disc jockey in 1927, playing 78 rpm records on the BBC.

Thanks to Stone's pioneering work, we can hear the difference between London streets of the 1940s and today. In the older soundscapes, vehicles are throatier, there are singing and whistling to be heard, and the voices of newspaper-sellers ring down the roads. Such recordings were until recently almost inaccessible, but the British Library has made them available once more. It also launched the UK Sound Map project in 2010, which collected recordings over a twelve-month period from anyone and everywhere.

13

A NEW BEGINNING?

In the 1950s, perhaps partly through the new-found recognition of the power of science in the Second World War, there was renewed interest in the effects of sounds on people. The effects of ultrasound and infrasound (sound lower than 20 Hz) were of particular interest. It had been discovered as long ago as the 1930s that, when bones are exposed to ultrasound, the marrow is heated more than the rest, but it was not until about 1950 that the first attempts were made to apply this to living subjects, using the heating effect to promote blood flow. Since then ultrasonic diathermy, as this technique is called, has been successfully used to increase the flexibility of tissues, reduce joint stiffness, and dull pain. Meanwhile, the distant ancestor of the medical imaging scans that everyone in many countries receives as an unborn baby was being trialled: a pulse echo system was used to measure both the thickness and the density of tissue. It also proved possible to use such systems to detect brain tumours. The fact that ultrasound travels readily in fluids but is absorbed over short distances in air means that the sound is naturally confined to the body, and is the reason for the use of gels between the ultrasound device and the skin, so that no ultrasound is lost by passing through air in between. The principle is the same as sonar: the waves are both delayed and modified on their journeys to and from the transducer, allowing both depths and densities of internal structures to be determined. The main difference is the frequencies involved: while military systems range from around 10 Hz for passive sonars and seismic sounders to 1 MHz for active sonars, diagnostic ultrasound systems usually operate between 1 and 10 MHz—or even higher in special cases, such as ophthalmic techniques. The reason is that, the higher the frequency, the greater the resolution (just as in astronomy, where radio wave images are intrinsically blurrier than those of light). At 10 MHz, for instance, the resolution is around one-eighth of a millimetre. Extremely high frequencies are not routinely used in diagnostics, since the amount of absorption by the tissues increases rapidly as frequency rises.

A new and unexpected use of ultrasound in medicine was also discovered at this time: in 1954 it was reported that the delivery of hydrocortisone to the hand joints of a patient suffering from arthritis was significantly enhanced if ultrasound was present: the energy of the sound waves being expended in increasing the rate of diffusion of the drug through the tissues.[1] This phenomenon is called phonophoresis (from the Greek for 'sonic transport').

Other work published in 1950 shows how mysterious the effects of ultrasound on the body can be, with evidence that the healing of bones was promoted by it.[2] Since then there have been claims and counter-claims, and no convincing mechanism has been agreed upon, but it seems unarguable that the effect is real—at least, sometimes. It is said that in some cases subjects fitted with non-functioning ultrasonic devices exhibited more rapid bone healing either than those with no such device fitted or than those with working ones!

At the other end of the spectrum, infrasound also has some odd effects. To begin with, the name is rather a misnomer—unlike ultrasound,[3] infrasound *can* be heard by humans, even at frequencies as low as 1 Hz, if the intensity is sufficiently high. However, no sense of a tone can be perceived; instead, there is a strange chugging, popping, or pumping sensation. Infrasound is all around us, arising from a whole range of causes. Whenever we walk, the up-and-down motion of our heads means that the ambient pressure around them fluctuates at our pace—around 2–6 Hz. Though the pressure variations involved are very small, so are those of quiet sounds within the audible range: typically around one-billionth of ambient pressure. So, walking generates infrasound waves at our ears at a level of about 86 dB. Ocean waves also fill our environment with fairly high-intensity infrasound at around 0.1–0.2 Hz —whether we live by the seaside or not.

Modern living exposes us to far higher levels of infrasound than this: simply driving along a motorway with the back windows open will generate very noticeable effects. The fluctuating, pulsing effect is unmistakeable and can be quite nauseating.

Because infrasonic wavelengths are similar to the dimensions of the body's organs, particular frequencies can cause disturbing vibrations of eyeballs and other structures. A less expected result is that infrasound causes drowsiness. In an experiment on a bus driver exposed to a controlled version of the actual infrasound he was likely to experience in his job (which was well below a noticeable level and had a frequency span of 1 Hz–20 Hz), drowsiness set in after just an hour or two, which is maybe something to think about on your next bus trip. When the infrasound was turned up to 120 dB (still an unnoticeable level), it was found that pulse rate and blood pressure both fell and even the alpha brain waves associated with wakefulness were replaced by the sleep-typical theta waves. Many church organs generate infrasound at just such levels, which may be some comfort to vicars with somnolent congregations. (But before switching off the offending organs, it might also be worth considering that such frequencies have also been found by some studies to generate feelings of awe and mystery.)

The normal audible range was also studied in the 1950s, and in particular the still rather mysterious cocktail party effect was scientifically reported for the first time. This effect does actually come in very handy at cocktail parties, allowing one to focus on a particular conversation when many others are going on—as far as the hearing system is concerned, this means classing everyone else's conversations as noise. It seems to involve comparing the split-second difference in arrival time between the ears and selecting the sounds made at the time interval that corresponds to the distance and angle at which the conversation one wants to hear is taking place. One of the effects of some forms of noise is the loss of this ability—even though the hearing range may be quite normal. This loss has a major impact on social life. The cocktail party effect has since been found to occur in many other species, including penguins, which use it to distinguish the cries of their offspring from the cacophony of similar cries arising from the huge crowds in which they cluster to keep warm.

In the early 1950s a number of lawsuits were successfully brought against industries in the states of Wisconsin and New York by workers who claimed that their hearing had been damaged by their working conditions, and that responsibility for their health at work lay with their employers. All but forgotten now, this 'damage to hearing scare', as it was known, was headline news and led to a more cautious attitude on the part of industry in the USA.[4] The use of the word 'scare' for what was actually a very reasonable concern shows that attitudes were still rather dismissive, but nevertheless noise control and regulation began to be taken more seriously by local and national governments. Following years of debate regarding the results of a Chicago noise survey carried out in 1947, an important new noise control measure was implemented, in the form of the Chicago Zoning Ordinance of 1957. Originally introduced to control industrial noise and later expanded to include vehicles, it was the first noise ordinance in the world to specify maximum noise levels.

Though ground-breaking and highly influential in legal terms, the Chicago ordinance was too simplistic in its approach to be used as the model for future such regulations. Many other factors than the measured level of noise are significant in deciding how much of a nuisance—how genuinely noisy—a factory really is. These factors include the time of day at which the noise is produced, the level of other noises from other sources, and the location: a factory that adds to the daily noise jumble of an industrial estate has a very different impact from an identical one operating at night in an otherwise quiet village, but the ordinance did not distinguish such cases.

Even before the Chicago ordinance had passed into law, a better alternative to the assessment of the significance of a noise source had been developed: the Composite Noise Rating (CNR). The CNR adjusted the threshold of acceptability of noise levels based on the level of background noise, the time of day and of year, the type of noise present (with tones and impulses being more annoying than broad band or continuous noises), and whether the type of noise was new to the community—the first time the fact that new noises are more annoying had been enshrined in law. The CNR was widely adopted but is now no longer used, having been supplanted by more sophisticated weighting systems.

Noise engineering

In 1952, the first serious attempts were made to make jet noise less problematic—but only for the planes.

It was clear by then that noise and vibration could cause costly damage to jet aircraft, significantly shortening their lives and reducing their efficiencies, and a technological solution was sought. The basic approach to these 'silencers', as they were rather optimistically called, was to make the high speed gases emerging from the jet mix with the surrounding air as rapidly as possible (see Fig. 31). This was first achieved by adding rows of toothlike projections to the inner edge of the end of the jet nozzle, and then by developing 'deeply corrugated' nozzles. Both these approaches increase the effective length of its perimeter and thereby extend the length of the interface with the air. Though a reduction of only a few dB was produced at the time, this type of approach was rapidly improved and adopted widely in passenger jets over the next decade.

CORRUGATED INTERNAL MIXER

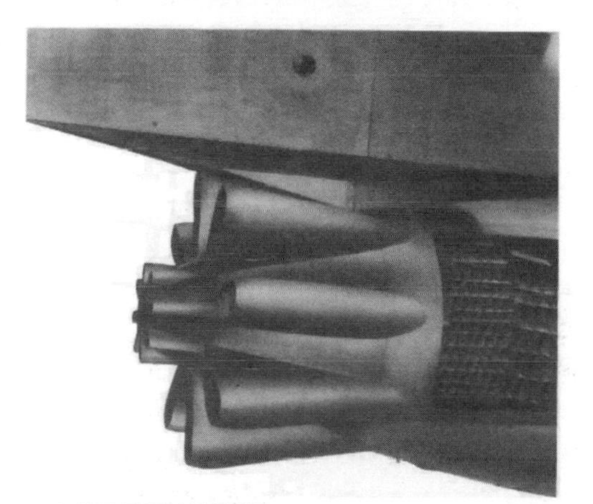

LOBE-TYPE NOZZLE

FIGURE 31. Why jet engines are no noisier: silencers.
Courtesy of College of Engineering, Purdue University, West Lafayette, IN.

In 1953, the first really successful application of active noise control was implemented by Harry F. Olson and Everett G. May, who wrote that their Electronic Sound Absorber 'consists of a microphone, amplifier and loudspeaker connected so that, for an incident sound, wave and sound pressure at the microphone is reduced. Thus it will be seen that the electronic sound absorber is a feedback system which operates to reduce the sound pressure in the vicinity of the microphone'.[5] The system soon became very popular in aircraft communication systems.

Despite the fact that noise was considered a problem by many powerful groups in the 1950s, and effective action was being taken in some areas as a result, the nature of the problem itself was still usually phrased in terms of the financial losses or physical damage caused. The annoyance caused to people was not in general taken very seriously, and cars in particular were often designed to be noisy—so long as it was the right kind of noise. For many, desirable cars were macho ones, so a roaring, powerful-sounding engine was what was required, not a quiet purring that would hardly inform people that you had a car at all—what would be the fun in that? Since then, this idea of designing cars that appeal to their prospective buyers through their sounds has become commonplace. Many millions have been spent on the study and then creation of just the right sort of clunkiness the owner of an expensive car wants when the doors close. The big issue here is again that of units. 'Clunkiness' doesn't sound—and isn't—a very scientific parameter, but then the calibration of the machismo of a door slam is not an easy task. Though a great deal of research has been put into the definition of such sound 'qualities', the area is rife with problems. Though there are many parameters on offer, from roughness to harshness and from shrillness to sharpness, none has enjoyed widespread acceptance, and it is still today only loudness that is a widely used measure of the subjective effect of noises.

However, cars in the 1950s did not always sound as they were designed to. Not for their owners the luxury of simply getting his or (much more rarely then) her car serviced every year or so and not looking under the bonnet in between. Just as with TVs and radios of the day, frequent attention was required—and that meant either expense, or getting your hands dirty. Hence, many car handbooks encouraged car-owners to take an interest in the performance of their cars and to assess it by listening to them, and the noises they made.

In Britain, the difficulty—not to mention hilarity—engendered by attempts to describe similar noises in the case of aircraft led to the facility used by Rolls-Royce to investigate problems with aero engines being officially designated the FSN Laboratory—FSN being short for 'Funny Sounding Noise'.

A nasty surprise was delivered to such noise campaigners as there were in the 1950s with the discovery that earplugs—the panacea offered to workmen in noisy conditions—were highly unpopular, thought of as unmanly and seldom used. This was particularly galling, since they had been issued as standard noise protection devices for some time, and yet no one had bothered to ask what the workers actually did with them. Industrial attitudes were clearly not much changed since Barr's boilermen.

14

THE POWER OF SCIENCE

In 1960, Leo Beranek, MIT professor and eminent acoustical consultant, claimed that noise control had been transformed from an art to a near-science in the previous twenty-five years.[1] But was he right? Were the people who tackled noise problems now going about their work scientifically? Certainly they *could* have been, as the understanding of acoustics in general and noise in particular was now quite sophisticated, thanks to the burst of research since the war. But it seems that not everyone had been reading the right books: like the factory-owner who came up with the brilliant scheme of spreading lots of microphones through his works, connecting them all—via an amplifier—to an array of loudspeakers on an adjacent piece of wasteland, and then turning it all on in the hope of 'sucking out' the noise. Other not very brilliant solutions tried at the time included installing sticky-backed absorbent foam, designed to line ventilation ducts, with the sticky (and not at all absorbent) layer outermost—because it gave a nice smooth airflow. Once the penny dropped that a nice smooth airflow was exactly what was not wanted, someone suitably junior was sent into the duct to pick the layer off with his (or her, maybe—history does not record) fingernails. And an actual *bona fide* engineer, trying to quieten a large factory building, tried filling just one of the many roof trusses with absorber to see whether it would be worth treating all of them. Not surprisingly, the minuscule reduction in overall noise was quite undetectable. Meanwhile, a widespread solution to noise problems in the 1960s, popular with builders, DIY addicts, and home-improvers alike, was the acoustic ceiling tile. They were not of much benefit in the first place unless used as part of the treatment of an entire room, and any slight help they might have been was frequently negated by the urge to cover them in thick layers of cheerful paint. Beranek himself mentions that, even in 1962, there was a widespread and 'deeply entrenched' belief that broken wine bottles beneath the stage were good for the acoustics of a concert hall.[2] In fact, the frequency with which such bottles are found is more likely due to the keenness of workers to hide their wine (bearing in mind that many such halls are in France and Italy). Another myth at the time was that a good way to test the acoustics of a hall was to drop a pin into a stiff hat (still very popular male headgear at the time) to see if the space was literally quiet enough for a listener at some distant point to hear a pin drop (or, rather, land). Such halls were reckoned—by some acousticians—to be 'perfect'. In fact 'terrible' would be a better word. If a quiet sound travels with unusual loudness to one point, it can only be because there is some favoured transmission path, which in turn means that listeners at other locations will not hear the sound and that music played in the hall will be heard very differently by audience members in different seats—exactly the reverse of 'perfection'. Such halls are really whispering galleries, and working conductors have long been concerned about them. As the great conductor Eugene Ormandy said to an enthusiastic architect: 'But I don't want to hear a pin drop, I want to hear the orchestra!'[3]

Of course, those are just isolated examples, showing that some factory-owners, architects, and engineers lacked basic knowledge of acoustics. One would not have expected anything similar of those employed on the sort of building where it is vital to get the acoustics right—like a purpose-built concert hall. Not only is its soundscape key to its success; it is also a place guaranteed to get plenty of publicity if it all goes wrong. So, the architects of the Philharmonic

Hall in New York employed the services of none other than Beranek. After making detailed studies of fifty-four other concert halls, he decided the solution was simple: play safe, by designing the hall like a shoe box. The architects, however, were much keener on a cool curvy 1960s kind of shape. And so that is what they built. Beranek, though doubtful, could not predict quite what the place would sound like, because the mathematical tools and computers required to work out the answer had not yet been developed, but he soon found out. When it was opened in 1962, everyone loved the shape but hated the noise: it sounded dead, bass sounds were weak, it was full of echoes—even the performers could not hear the performance properly. It was too reverberant, too harsh, too mushy—people couldn't agree on quite what the problem was, except that there was one. Arguments, criticism, and blame went on for decades to the severe detriment of many a reputation, including Beranek's. Finally, after years of struggle with makeshift solutions, the whole interior was ripped out and remodelled at great expense, reopening in 1976 as a much more acceptable acoustic space (see Fig. 32). It was even renamed the Avery Fisher Hall in the

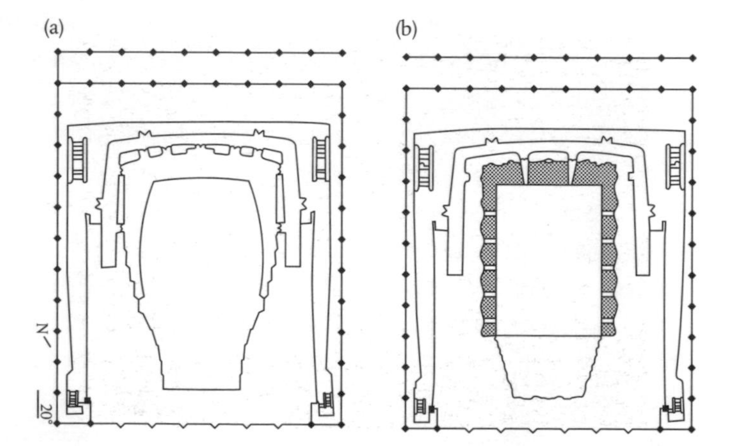

FIGURE 32. The Philharmonic Hall, in its original and renamed, remodelled, versions.

hope of drawing a veil over its troubled youth (see Fig. 33). Its new shape closely approximated the shape of the Symphony Hall in Boston—that of a shoe box.[4]

Following this humbling experience, Beranek and others realized that noise control engineering had not *quite* reached the required state of scientific perfection, and there was a brief frenzy of measuring the shapes and sounds of concert halls, from which did emerge a set of serviceable rules as to how to get it right. Or at least, not terribly wrong.

An enormous influence on the process of building quieter spaces was the rise of the electronic digital computer. In particular, a 1966 project carried out by Bell laboratories in the USA was groundbreaking in using the newfangled gadgets to analyse digital versions of tape-recordings of sound pressure levels at different points in a number of rooms, to relate reverberation to room dimensions.

Just as in the negative effects of sound on people, so in this case of the positive effects there was a clear need to link physical parameters such as reverberation time to some subjective—but well-defined—quality. Beranek succeeded in tracking down one

FIGURE 33. The Avery Fisher Hall interior.

key link when he managed to relate the difference in arrival times between the direct sound and the first reflection of it to measures of listening appreciation. The new-found enthusiasm for making and processing huge numbers of measurements of 'good' acoustic spaces to augment existing theoretical predictions and rules of thumb also provided quantitative confirmation of the widespread belief that older concert halls sound mellower. The reason is that their surfaces tend to be irregular (and also the unsatisfactory old ones are more likely to have been knocked down or turned into bingo halls).

Another small first for the 1960s was the active encouragement by rock stars for their fans to do their bit for noise pollution—by adding to it. The first band in this proud tradition was the Rolling Stones, who, on their 1969 album (or, as they were called then, LP), *Let It Bleed*, included the suggestion that 'This Record Should Be Played Loud'. Mick Jagger had presumably not been convinced by the statement that same year by UNESCO's International Music Council that proclaimed 'the right of everyone to silence, because of the abusive use, in private and public places, of recorded or broadcast music'. Not wishing to be left out, several other bands followed in this new-found approach, perhaps most famously David Bowie, who insisted that *The Rise and Fall of Ziggy Stardust and the Spiders from Mars* (1972) was 'TO BE PLAYED AT MAXIMUM VOLUME', the capitalization presumably representing a shout, as it is often now interpreted in emails. In a touching show of conscience on the part of Bowie or his record company, his instruction was removed when the record was reissued in 1999. Not that all band members were so keen on noise: Noel Redding, of the Jimi Hendrix Experience, even lent his image to a Noise Abatement Society poster (see Fig. 34).

A possibly ideal solution to the problem of making records that would be noisy only for those who wanted them to be was provided by the Beatles. In their 1967 track 'A Day in the Life', they included a note of around 15 kHz—too high for older people (that is, anyone over 30) to hear but within the frequency range of teenagers who

FIGURE 34. Noise Abatement Society poster, late 1960s.

had not yet been to too many pop concerts. It does, however, apparently annoy dogs. (A more deliberate attempt to engage canine listeners was made in 2012, with the world's first advert to include ultrasounds that dogs, but not their owners, could hear. The advert was, unsurprisingly, for dog food).

Meanwhile, some real science was at last being done on how noisy a noise actually is: how the subjective experience of its unpleasantness relates to the measurement of its sound waves. As arranged by NPL, following up on their successful jury approach to motor horn assessment, fifty-seven brave volunteers wrapped themselves up against the nasty weather of autumn 1960 and sat patiently by the side of a test track, writing down just how annoyed they were while nineteen vehicles drove past them six times each.

John Connell

Noise was of particular concern to Londoners and New Yorkers in the early 1960s because of the expansion of transatlantic air travel. A major contributor to this growth was American Airlines. It started a transatlantic jet service based on the Boeing 707, which cut transit time between the noisiest cities in the new and old worlds in half. At the same time, the number of seats available doubled, allowing an economy class to be introduced. The prospect of cheap fast travel led to a great surge in demand, until in 1958 more people crossed the Atlantic by air than by sea. And a large proportion of them were travelling between Heathrow (then called London Airport) and New York's John F. Kennedy airport (then called Idlewild), which grew rapidly over the next few years.

Jet planes like the Boeing 707 were not only large and powerful. The frequencies of the noise they made were closer to peak hearing sensitivities too. In terms of overall sound level, while propeller planes had generated around 85 to 90 dBA at take-off when heard 300 metres away, the new jets were producing between 100 and 105 dBA under the same conditions. With jets, the noise level also varies greatly depending on the speed of the jet flow, which itself is annoying. In fact, the sound power output is proportional to the 8th power of this speed, so doubling the speed of the output results in an increase in the sound power by a factor of 256.[5] In consequence of all this, a steep rise in noise complaints was recorded.

What the new but noisy 1960s needed was a new hero—and it got one, the first anti-noise campaigner to catch the UK public's attention since Babbage, and in a considerably more positive way. And his work had a far more lasting impact.

John Connell was a businessman, and his story really started in the long hot summer of 1959 when he noticed the large number of letters in the newspapers about noise—peaking as usual in the summer months when windows are open and planes fly low (because the air is hotter and hence thinner, they must descend to thicker areas closer to the ground). As a result, Connell formed the Noise Abatement Society (NAS), a replacement for the earlier Noise Abatement League, which had been formed in the 1930s but had closed because of lack of funding in 1951 without having accomplished much (see Fig 35).

To test the water, Connell placed notices in several newspapers asking anyone concerned about noise to write to him. He received over 3,000 replies. Over the next few years, the letters continued to arrive on his doorstep by the sack load: at last, people had someone to whom they could write about the problem that was blighting their lives, in which the government seemed completely uninterested, and about which few who did not suffer from it were bothered.

There was a general election in 1959, and one of the NAS's first significant actions was to write to all 1,564 parliamentary candidates, asking them to push for new legislation against noise. All but three said they would. And what is more, they did: with MPs fearful of the sudden rise in public interest (and, who knows, maybe rather annoyed themselves by noise), the *Noise Act* was rushed through and appeared on the statute books the following year. There was just one problem: it stepped back from dealing with the rising issue of aircraft noise, by excluding it from prosecution as a nuisance—simply because to do so would have crippled the burgeoning airline industry.

That same year, Conservative aviation minister Duncan Sandys allowed night flights from Heathrow, and, in response, Connell took direct action, turning up at 2 a.m. on Sandys' doorstep.

FIGURE 35. John Connell OBE, founder of the Noise Abatement Society, a charity in the United Kingdom.

Sandys's reply to Connell's queries about the Heathrow noise was not very encouraging—noise, he said, was an unavoidable fact of modern life near an airport. There was simply nothing to be done about it.

Fortunately, there was. As a result of the publicity generated by Connell, night flights from Heathrow were halted. Furthermore, his work had raised interest in the noise issue, and in 1960 the government asked Sir Alan Wilson to convene a committee to 'examine the nature, sources and effects of the problem of noise and to advise what further measures can be taken to mitigate it'. The Wilson Committee's remit extended over all major noise types, including motor vehicles, buildings, railways, aircraft, industry, construction sites, and entertainment activities. It also reviewed the current jumble of laws on the topic and compared noise in the city with that in the country. The committee's report, published in 1963, was the definitive work on noise in the UK for many years and led to many of today's approaches to its measurement, analysis, and mitigation. The committee relied on a wide range of data-collection methods, from surveys and experiments to literature reviews and consultancies. Among many other things, it developed noise contour maps (initially around airports). These are now the standard way of representing noise level over geographical areas. In its wide remit and its level of funding, the Wilson Committee was groundbreaking. It was really the first national attempt to tackle noise.

A committee for peace

The Wilson Committee wished to conduct an 'experiment on the level of noise which is acceptable' and it asked NPL to carry it out. The results were to have considerable importance for future work: although aircraft generated a wider range of noise levels than did cars, the range of people's responses was the same in each case—which is really quite surprising. It means that, subconsciously, listeners must classify the type of noise in terms of some kind of range and scale their answers accordingly. So, when they decide something is 'quite noisy', they mean 'quite noisy—for a plane'. Clearly, this is a real problem if one wants a simple way of relating noise level to perceived loudness, irrespective of the thing that is making the noise. In fact it just cannot be done.

The Wilson Committee's report was issued in 1963, and its chief conclusion was that the main aim of noise legislation should be prevention, rather than compensation, as had often been the case previously.

On neighbourhood noise, the Wilson Committee was probably least successful. Its conclusion there went little further than advising people to talk to their neighbours instead of listening to their noise—which was not a very British thing to do. In fact, the main way to deal with the large proportion of noise that was caused by thoughtless and irresponsible behaviour was education—just as it had been for the previous generation and the one before that. As the report put it, in a way unlikely to inspire much hope in the reader: 'A substantial amount of annoyance [due to noise in towns] is caused by thoughtlessness and carelessness. The most effective remedy for this would be a general improvement in good manners.'[6]

In general, though, the report's many recommendations were eminently sensible and practical, from smoothing traffic flow to training architects in the principles of sound insulation. Motor horns should be forbidden at all times in built-up areas. Perhaps most importantly, the report set suggested specific limits both for acceptable levels of noise (in living rooms, for instance) and for the noise output of new vehicles. The Heathrow noise problem was cited as one of the most difficult in the world. The suggestion that aircraft could be towed around by tractors rather than taxiing was considered but was rejected as too inconvenient. The main response was the proposal for government grants for noise insulation of dwellings near Heathrow—a measure that was adopted. As an aside, the report also noted the possibility of the development of quieter aircraft, but was not hopeful on this score since, by the 1960s, the use of silencers on planes was making little headway because of its impact on reducing the thrust of the plane and hence its efficiency. However, a new approach, soon to be universally adopted, was that of the turbo-fan engine, which is designed in such a way that only a small fraction of the air drawn in by the engine passes through the combustion chambers to be squirted out as a high-speed jet. Most of it is directed along the engine's exterior, creating a buffer zone between the air jet and the surrounding air. The result is a smoother mixing of the jet output with the surrounding air, leading to less turbulence and significantly less noise.

The report also revealed that the UK public were getting increasingly irritated by noise, with 91 per cent being conscious of noise outdoors compared with only 42 per cent in 1948. Meanwhile, 50 per cent were now disturbed by that noise compared to 23 per cent previously. Traffic noise was the most bothersome. However, there was a modest reduction in the proportion of people who noticed or who were annoyed by indoor noise, perhaps as a result of better building techniques.

There is little doubt that the report was quite right to suggest that more understanding of noise would be a good thing. In fact, it was all that would have been needed to stop the first legal case involving noise in modern times from falling through. The case was that of a workman who, in 1966, had been employed on a construction site at Essex University. His job was to fix sheets of metal to concrete window lintels, which he did by banging in around 130 rivets per day by means of a cartridge-assisted hammer, which was held near his right ear. After two weeks' work he became first dizzy and then completely and permanently deaf in that ear. His employers had failed to provide him with any protection, but such was the general lack of noise awareness at the time that, when the case was heard in 1969, the judge found that the employers were not negligent in failing to appreciate that the noise of the gun was a hazard: it was agreed that no one could reasonably have expected them to know such an arcane thing.[7]

In addition to the fact that people judged noise levels partly from their knowledge of the source, it was found that people weight the intrusiveness of noise depending on whether they are indoors or out—they are more tolerant of noises outdoors, by a very significant factor: 18 dBA. In other words, a noise heard outside has to be sixty-three times as powerful as one indoors to be perceived as equally intrusive. And, if a dwelling does not provide a noise reduction of at least 18 dBA, then external noises are more annoying indoors even though they are quieter.

The finding generated the uneasy feeling that the whole issue was so subjective as to be impossible to pin down. However, consistent results were given by many experiments and tests, and today the indoor/outdoor dichotomy is well accepted.

Attempts were made to compare the findings of the Wilson report with the results of survey and studies in other countries, but several problems emerged. Words like 'annoyance' do not always have exact equivalents, and survey questions and scoring systems differed.

An eight-nation study revealed a more fundamental problem with the annoyance response.[8] Noises in the real world that are referred to as annoying elicit feelings of both disturbance and anger because of the feeling of powerlessness. This—though rather obvious after it has been pointed out—is a problem, because it casts doubt on the comparison of such real situations with laboratory studies, which do not usually anger their subjects, just disturb them.

Furthermore, just as in the cases of the Railway Bonus and the Aircraft Malus, the listener's view of the danger associated with the noise and of the necessity for making it affects his or her response. Yet another factor is the degree of trust the listener has in those who are believed to be responsible for the noise. For example, surveys of people living near Dusseldorf airport have consistently shown higher levels of annoyance than those of people living near other comparable airports.[9] The difference stems from 1992, when the local authority, which had previously owned the airport, sold it to a private company, which then launched an expensive and successful campaign to get permission for a 30 per cent increase in flights. Similar increases in annoyance are recorded among those who believe that noise is damaging their health compared to those who have no such belief.[10]

All this makes prediction of the amount of annoyance that noise will cause very complicated, and turns the comparison of laboratory studies with reality into a minefield—but, as they say on all the best management courses, problems are just solutions in disguise. These sorts of insights into noise show just how essential it is for those who are—or are perceived to be—responsible for noisy activities to explain clearly the reason for them and to engage with those affected. The noise itself should never be its own message.

However, the Wilson Committee did come up with what seemed to be an adequate way to relate measurements to subjective reactions, the Noise and Number Index (NNI). The NNI attempted to quantify the subjective noisiness of aircraft, and it used both measured sound levels and the number of aircraft per day (or night) as a key annoyance factor. NNI contours were produced for Heathrow and subsequently for other airports too.

It was what many had been waiting for: the NNI was rapidly pressed into use to determine which of those homeowners who lived near Heathrow should qualify for an improvement grant, and was also used for specifying land use near airports—that is, yet again, zoning. Unfortunately, however, the sound levels were to be measured not in dBA but in PNdB (perceived noise level in decibels), a unit with which contemporary technology was not equipped to deal—so officials simply measured in dBA and added 13, which was not very likely to give the right answer. Nevertheless, one great benefit of the system was that automatic noise-measuring devices were set up along flight paths—despite the facts that they measured the wrong units, often failed owing to corrosion, and were cowled to protect them from rain and wind in ways that added great uncertainty to their results. But, luckily, they were easily visible to pilots, who treated them with great respect and hence kept very closely to the agreed flight paths. This may have done more to protect people from noise than any other aspect of the scheme.

While Connell's solution to the Heathrow problem, which was to move the airport to Foulness,[11] was not seconded by the Wilson Committee, his efforts and those of other members of his society were successful in influencing other aspects of the report, such as the introduction of rubber dustbin lids and plastic milk crates (see Fig. 36)—which really did improve somewhat the mornings of millions. In fact, both through the direct work of the NAS and through his role in triggering the government's decision to set up the Wilson Committee, John Connell deserves to be far better remembered than he is—as perhaps the single most important figure in the history of the fight against noise in the UK.

FIGURE 36. John Connell OBE, with one of his highly successful rubber-lidded dustbins.

International disharmony

By now, it was abundantly clear to many people in many countries that noise was a common issue and, therefore, especially in the area of aircraft noise, that a unified approach was required. So all that was required was for the rest of the world to start using the same approaches and units and everything would be fine. Except the rest of the world didn't want to. Though everyone wanted an internationally consistent approach, each of the major players—the USA, Germany, the UK, and the Netherlands—wanted something different. In fact, each thought its own unique approach was clearly the only way to go and sometimes could hardly be induced to take the others seriously at all.

The reason for this disharmony is that the differences between the approaches did not arise from technical questions like which weighting factor worked best or how to set up the equipment—rather, each country's proposal was based partly on the types of noise that it found most problematic and partly on its particular research traditions.

German acousticians disliked the NNI because it excluded the impact of the duration of noise: their laboratory- and studio-based research into industrial noise, which was much more extensive than that made in other countries, suggested the importance of this.

In the Netherlands, a wealth of surveys led to policies favouring zoning and surveillance. Dutch acousticians found the measurement of noise durations impracticable and preferred simple dBA units as a result.

Noise experts in the USA had recently made extensive studies of the annoyance caused by jet noise compared to that of propeller aircraft, the results of which had led to the development of the PNdB scale.

UK acousticians liked the NNI and the jury approach and feared the public's distrust of complex science-based standards.

Germany was not impressed by this view, arguing that it was illogical. The Dutch thought the US approach was far too inaccurate and that the German approach was not practical, and the USA thought the Dutch and Germans were unnecessarily exacting. The British thought all the other three countries depended too much on objective measurements. And so on.

For the UK, though, there were local difficulties too. A survey of annoyance near Heathrow (carried out in 1967 and published in 1971) cast doubt on the validity of the much-loved NNI. The relation between the number of flights and the annoyance caused by them was not the same as the Wilson Committee had decided, probably because, in the 1960 work, there were several values of plane number for which hardly any data had been collected. The new report also noted that social attitudes towards noise had hardened in the period between the new study and the 1961 one.

Fortunately, the government had set up the Noise Advisory Council in 1970, so the Minister of Trade asked it what to do. Nothing at all, came the rapid response. After all, a modification might 'bring into disrepute the entire basis of such methods for quantifying nuisance from aircraft noise'.[12] So, nothing much more was said, and the NNI continued to be used for the next two decades, until it was replaced by LAeq. LAeq is a measure of the total energy of a period over which noise is present—it is the continuous sound that would contain the same amount of energy as the actual sound present. It was found to be well correlated with the amount of disturbance and is still in common use in the twenty-first century. UK planning applications, for example, are assessed in terms of this measure.

In the USA, the fight against noise was going through a similar process to that in the UK in the previous century, with campaigns by pressure groups and vociferous individuals, all mostly focused on a single issue. And that issue was airports. There was little organized tackling of wider noise issues until 1969 with the passing of the National Environmental Policy Act (NEPA), augmented by the Noise Control Act (NCA) in 1972. The Acts were the result of the efforts of the US Environmental Protection Agency, and in particular its warning to Congress that thirty million US citizens were exposed to noise high enough to cause hearing loss—leading to the inclusion in the NCA of the statement that it was the USA's policy 'to promote an environment for all Americans free from noise that jeopardizes their health or welfare'.[13]

Bolstered by the Quiet Communities Act (1978), the Environmental Protection Agency regulated many noise sources, introduced noise labelling of products and funded noise research. It determined that long-term noise levels in excess of 55 dBA were sufficient to damage health, and estimated that almost 100 million US citizens were exposed to long-terms levels above this value—a startling figure, given that the population numbered only 213.9 million at the time.

The fight against noise in the USA was, however, soon to be hamstrung. Recognizing the scale of the problem, the government decided that the best way to tackle it was through a 'total national program' to be delivered through cooperation between the many relevant agencies.[14] Unfortunately, with no agency in overall charge and a marked unwillingness for people to work together, it soon became clear that the total national programme was neither total, nor national. In fact it was not really much of a programme. Such problems remain to this day—the agencies are not well coordinated and do not even agree on what noise levels or types are significant, let alone how to deal with them in a unified way.

As a result, while the USA put in place policies and regulations that were the most advanced in the world from the late 1960s until the late 1970s, their practical legacies were quite limited. The most significant successes were in the area of aircraft noise. Not only were domestic aircraft noise levels reduced but, even more importantly in the long run, a significant degree of international cooperation was achieved. In particular, in 1972 an agreement was reached between the International Civil Aviation Organization and the US

Federal Aviation Administration that all new aircraft to be flown in Europe or the USA had to meet certain certified noise levels. Other countries soon followed suit. This has been a highly effective measure, resulting in dramatic reductions in both noise levels and numbers of complaints from many of the world's airports.

The rise of the soundscape

By the 1960s it had become clear that the full measurement of the sounds of even a small urban area required a vast number of instruments at many locations, collecting sounds over a long period and then analysing them in great detail. And even then the effects on the inhabitants would not be certain. Fortunately, there was another way of thinking about environmental noise and its effects: the soundscape. Defined and developed by Murray Schafer in his 1977 book *The New Soundscape*, a soundscape refers to the whole sound environment in which the hearer exists.[15] Typically, there is some 'keynote' sound, which helps to define the type of place, such as the roar of traffic or the sound of the sea, together with sounds explicitly used for signalling—from words to car alarms—which are the parts of the soundscape of which people are usually aware. There are also soundmarks, analogues to landmarks, which help structure the space. These might include the ticking of a clock, the playing of a fountain, or the squeak of a gate. In contrast to modern urban soundscapes (which he deplores), Schafer describes the spontaneously melodic soundscape of Skruv, a tiny village in Sweden with a few hundred inhabitants. He found that, by a remarkable coincidence, the street lighting, electric signs, and generators made sounds that together produced a G-sharp major triad, and that the F-sharp whistles of passing trains developed this into an intermittent dominant seventh chord. Whether the inhabitants appreciated their good fortune was not mentioned.

One of the most interesting and practical corollaries of this perspective is that areas affected by environmental noise might be improved by adding more sound to them, rather than by removing it. For example, a fountain might be added to a city square, or the sound of the wind enhanced by a belt of trees.

In Schafer's view, the idea of a soundscape that could be moulded by its inhabitants was very timely, as he believed that the time had come when individuals were about to wrest power from industry, just as industry had done from the church in previous centuries—a topical view in the 1960s. And in the area of aircraft noise, the public were indeed starting to take matters into their own hands, even if they were not yet designing any soundscapes—except, perhaps, in their gardens. It would not be long before wind-chimes and water features would indeed be modifying the sounds of private gardens.

More recently, some have suggested that functional sounds could be made more tuneful—for example, by changing the noises of car horns to more musical ones. Of course, this is exactly what has happened with mobile phone ring tones—no longer are they just boring buzzes but a whole range of tunes, songs, even witty remarks and 'special effect' noises. Does this improve the soundscape? You be the judge.

The startling story of Concorde

In 1967, the roll-out of the first Concorde marked the fulfilment of a dream that had begun decades earlier. People fell over each other in their attempts to praise it. Morien Morgan of the Royal Aircraft Establishment even declared: 'If God meant airplanes to fly, he meant them to be this shape.'

Concorde was indeed an amazing achievement, with groundbreaking digital control systems, the first computer used in a commercial aircraft, fly-by-wire technology, and braking mechanisms not introduced into other planes until thirty years later (see Fig. 37). The price tag was amazing too—£1.134 billion, with even the pilot's seat costing £80,000,000. People loved all sorts of things about it—the new shape, the scale of the project, and the fact that British and

FIGURE 37. Concorde roll-out at Toulouse, 1957.
Hulton Archive/Getty Images.

French teams had developed it together—despite, that is, agreeing on almost nothing about it: even Concorde's final 'e' was a source of bitter argument. Originally, in both the UK and France, 'Concorde' was the accepted version, but the then Prime Minister Harold Macmillan removed the final 'e' as a result of a dispute with his opposite number, Charles de Gaulle. Tony Benn (the Technology Minister) put the 'e' back again, much to the irritation of the nation (according to the press).

But what people loved best, and talked about most, was the speed—600 metres a second, faster than a rifle bullet. It was not just fast; it was *supersonic*.

And that was the problem. If any object is moving, the sound waves from its front end travel at around 1,200 kilometres per hour through the air, which is the speed of sound.[16] But if the object is moving through that same air, it starts to catch up with the sound waves, and the sound waves start to catch up with each other too.

If the object reaches the speed of sound, it succeeds in catching up with the sound waves; so they then all catch up with each other, and, instead of the energy they contain being spread freely through the air, it all piles up in the same place. The concentration of energy is called a shock, and it sounds like an explosion—what is now called a sonic boom.

Usually, when there is an approaching plane, there is a slow crescendo of sound. But with a supersonic fly-over, there is a piling-up of sound waves. So the plane passes over first, and then all the sound arrives at once. There is, in fact, a series of booms, all close together, which are produced by different parts of the plane, but the loudest are those produced by the nose and the end of the tail. So what is heard is a brief, very loud, series of noises (see Figs 38 and 39).

Concorde's sonic boom, though startling, was not unexpected. In fact, some people were quite excited about it, since it marked, for all to hear, the point at which the aircraft broke the 'sound barrier' and started to travel supersonically. Before the first flights, scientists speculated about the effect of the sound, asking questions like 'Will the short duration of the phenomenon make the annoyance insignificant?'[17] and 'What level of bang will the public tolerate?'[18] They were soon to find out that the answers were 'No' and 'None at all', which is hardly surprising: brief bursts of noise are far more annoying than sounds that take their time.

There were plenty of complaints about Concorde in the UK in the 1970s, and at least one fully-fledged home-grown campaign, the UK Anti-Concorde Project, led by a retired teacher called Richard Wiggs. Though it adopted techniques that were innovative for the time, like taking out anti-Concorde adverts in newspapers and giving away multi-coloured anti-Concorde T-shirts, the project had relatively little effect on the public, who loved the speed and the look of the plane, or on the government, who had spent vast sums on it. It was the people living near John F. Kennedy airport in New York who organized themselves into a protest movement with international impact.

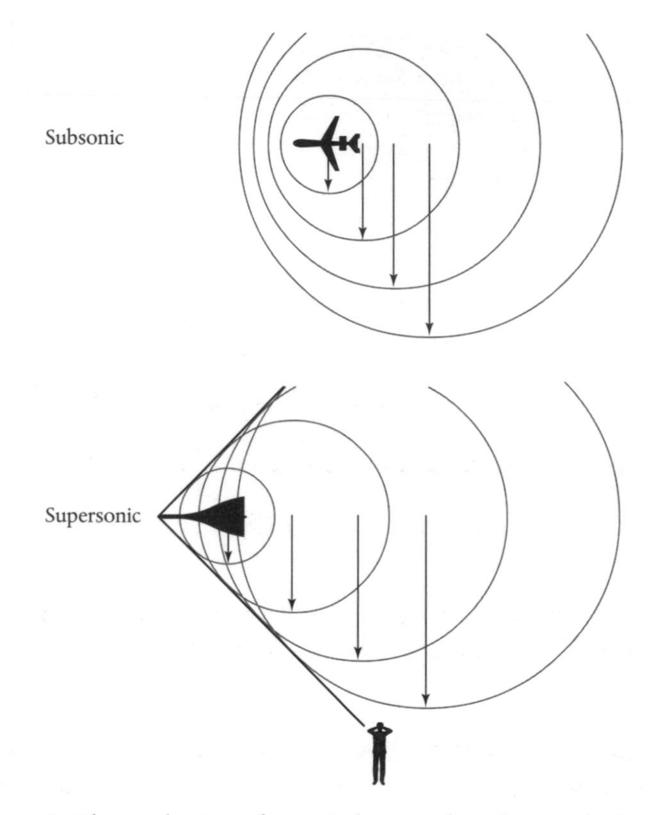

FIGURE 38. The production of a sonic boom: when the speed of sound is exceeded, the overlapping wave fronts form a shock front.

Redrawn from Rupert Taylor, *Noise* (Harmondsworth: Penguin, 1970), 106–7, figs. 28 and 29. Copyright holder not traced.

Among many other things, the protestors threatened to block roads around JFK, claiming that it would take Concorde's passengers longer to leave the airport than to cross the Atlantic. As a result, in January 1976 the US Congress banned all Concorde landings in the USA. Though this was speedily overturned by the Secretary of Transportation, flights were still permitted only to

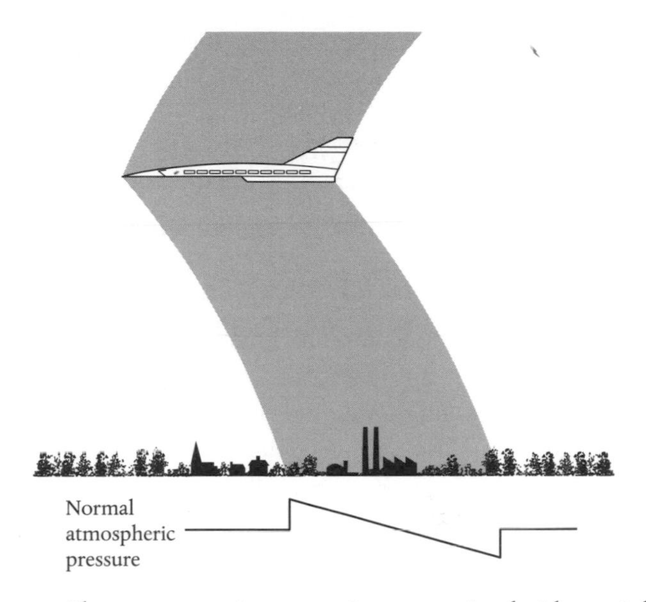

FIGURE 39. There are two main pressure jumps associated with a sonic boom. Each leading and trailing edge of the fuselage also produces a smaller such jump.

Washington Dulles airport. The ban on flights to JFK was finally lifted in October 1977, but by then the publicity that the campaigners had attracted had sealed the fate of Concorde overseas: many countries simply said no.

The Indian government allowed Concorde to access the subcontinent, but only at subsonic speeds, which meant the route was never viable. As Ken Binning, Concorde's Director-General, recalled: 'I had a very uncomfortable morning in Delhi, in front of the Secretary for Transportation, attempting to explain why it was perfectly sensible to over-fly six million Indians supersonically and wake them up in the middle of the night, but we didn't do the same thing to Europeans. "Interesting," he said, "is this because there is some biological difference, in your view, between Indians and Europeans? "'[19]

Even countries that had not been put off Concorde by the bad publicity were soon deterred by the thing itself: in Malaysia, supersonic flights were allowed—exactly three of them, after which noise complaints led to an outright ban.

The creators had dreamed of an international web of supersonic services linking distant parts of the world with flights that took a few hours or less; instead, Concorde ended up with routes connecting London and Paris to just three countries: the USA, Bahrain, and Brazil. Even these were hedged in by bans on over-flying certain territories, and speed restrictions over others, which increased flying times and dented profits. As a result, Concorde lost money more or less continuously for the rest of its career. Finally, in 2003, a Concorde crashed in Paris, killing all its crew and passengers, and the remainder of the fleet was grounded soon after—permanently.

As mentioned earlier, there is actually a lot that can be done to make the engines of planes quieter. But supersonic planes would still make sonic booms, even if their engines were completely silent. That point is, however, academic in the case of Concorde—it made a phenomenal noise, however slowly it flew. In fact, it was probably the noisiest commercial airliner ever, at any speed. Its whole shape and structure was designed to fly efficiently only at supersonic speeds, and, at the humble speeds of around 800 kilometres per hour of more conventional craft, it was very loud indeed.

The fact that Concorde was granted an airworthiness certificate at all was a credit to the power of political will over scientific rigour. After it had been obtained, many people under the flight path wanted to know why, but the response to a Parliamentary Question was that the noise-monitoring equipment at Heathrow had measured acceptable noise levels. A fly-over of the Farnborough Air Show was also rather less ear-shattering than expected, and the negative publicity became slightly less vociferous—since it was only later revealed that the noise-monitoring equipment had been 'mis-set' and the fly-over had been made by a Concorde from which all the seats, most of the fuel, and just about everything else had been removed.

The anti-Concorde protests were the start of worldwide public concern about aircraft noise. Before then, people had been annoyed;

now they were annoyed and vocal—and politicians started listening. As a direct result of the JFK protestors, noise control regulations were applied to aircraft flying out of their airport. In France, acoustic barriers were introduced to fend off growing local criticism, and in the UK, the large-scale production of Wilson Committee-type noise maps (rather optimistically called tranquillity maps for a while) of airports began.

For people living near London, the air became suddenly quieter after Concorde's final departure from Heathrow on 23 October 2003.

15

A NEW POLLUTION

It is 1970, the sky is red, people are dying, and it is all the fault of jet noise—at least, that is, according to the BBC's *Doomwatch*, a series of gloomy science-fiction tales with a new Earth-threatening topic each week. The roping-in of aircraft noise along with more conventional science-fiction menaces such as killer mutant rats and superbugs shows, however, just how much the subject had become part of the public mindset. Concern about noise was spreading, in part because of the very-well-publicized issue of sonic booms, itself raised by the fame of Concorde. On this basis, a large number of papers, books, and TV discussions of the subject appeared throughout the 1970s in both the UK and the USA, and groups and individuals who had previously been regarded with scepticism and suspicion were gradually accorded more respect—and airtime. The groups themselves were changing, too: in the 1960s, with a few notable exceptions like the NAS, noise action groups tended to have a fairly narrow focus and campaigns tended to be of the NIMBY (Not In My BackYard) variety, but now their comments and complaints became more general and referred to the effects of noise, not just on themselves, but on the environment. At the same time, many other environmental issues were also being discussed, including such topics as animal rights and pollution of all types. It is this new approach to the world's problems in the 1970s that led to noise being added to the list of pollutants.

By now, European countries were catching up with the USA in introducing national policies regarding noise. The expansion and development of the European Economic Community (EEC) gave rise to a renewed hope for uniform international laws. The idea of setting levels within countries for particular sources and situations was increasingly adopted in several nations. In the UK, the result of government legislation and public interest, thanks largely to Connell, was that earth mounds and barriers were added beside existing roads, and noise barriers were constructed round airports.

Working rather against the general theme of noise reduction in 1970, inventor Crawford Senior was doing his bit for noise by inventing special gloves made of metal or other hard materials and intended to increase the sound of applause—an excellent example of a solution looking for a problem.

Noisier or nicer?

One thing that had now become clear, following the determined but failed attempts to reach international agreement on almost any aspect of noise pollution in the previous decade, was that noise is irretrievably caught up with the differing views of individuals, such that, looking back on it now, the efforts to determine a universal level of tolerable noise seem hopelessly naive. It had failed in the 1950s, again in the 1960s, and once more in the 1970s. In the twenty-first century people have pretty much given up on it. On the other hand, more and more studies and surveys were making it widely understood that physiological damage due to noise, hitherto a major focus of discussion, was actually very largely a red herring. Of far more relevance in terms of the numbers of people affected was the impact of noise levels on what is now referred to as quality of life, through sleep disturbance and annoyance. And this, in turn, meant that far lower levels of noise were acceptable, broadening enormously the scope of the problem. It was clear that new ways of assessing and dealing with noise were required.

A modification of Schafer's idea of the planned manipulation of the soundscape by adding to it was provided from the world of music, in the form of ambient music. The most famous early example is Brian Eno's 1978 album *Ambient 1: Music for Airports*. In a move reminiscent of the early noise musicians of the 1900s, it even included a manifesto: 'Ambient Music must be able to accommodate many levels of listening attention without enforcing one in particular; it must be as ignorable as it is interesting.' While this to some extent followed up Eric Satie's furniture music, it was much more successful—and not just in airports. Soon, similar music was being played in dentists' waiting rooms, lifts, shopping malls, and finally supermarkets, where now it is of course very common—and equally, of course, not to everyone's taste.

In the first half of the twentieth century a great deal was done to quieten some noises at source, in particular by the introduction of mufflers into cars, the banning of motor horns in many times and places, and the introduction of bypass engines into planes. The results were impressive: modern jets produce less than 0.02 per cent of their output as sounds, and cars are even better, producing only about 0.001 per cent. As a result, by the 1970s there was little more to be done in these areas. However, despite the growing interest in the noise problem and the fact that building acoustics were now very well understood thanks partly to the increasing use of computers, it is a sad fact that most of the buildings constructed in the 1970s are very little, if at all, quieter than those built previously. Flats in particular were often built with quite inadequate party walls, and no double glazing. Effective construction techniques and materials that would have considerably lessened noise were now available, and yet neither builders nor planners took much interest in them. As a result, Rupert Taylor, a noted noise commentator and consultant on noise, complained in 1970 that 'design engineers, with some notable exceptions, even if they do have some knowledge of acoustics, forget all about it'.[1] No doubt another issue is simply that noise-reduction constructions would have added to the building cost, so, until either regulations forced the builders to take noise into account (as is now the case in many countries) or there was real interest from buyers in such an idea, nothing was likely to happen.

Since houses were no better protected than before, all that could be done for their inhabitants was to add noise-control measures after the fact: the government insulation grants that had been provided for dwellings blighted by the noise from nearby airports as a result of the Wilson report led to thousands of homes being insulated during the 1970s. By the early twenty-first century, the insulation of about 50,000 UK dwellings had been funded in this way.[2]

A new and sinister arena for the use of noise was Northern Ireland, where, in the early 1970s, Operation 'Demetrius' involved the use of white noise[3] against IRA suspects prior to interrogations by the Royal Ulster Constabulary, who were acting on guidance from the government. Noise was one of five newly recognized interrogation techniques, along with being made to stand in stress positions for long periods, hooding, sleep deprivation, and lack of food and drink. In November 1971, once the use of the techniques had become public knowledge and led to widespread concern, the UK government commissioned an inquiry and found the five techniques to be illegal. They were banned the next year.[4]

The underwater soundscape

Underwater acoustics research had continued steadily since its resumption during the Second World War. The primary reason for this was the onset of the cold war, which redrew the map of underwater military action. The development of nuclear submarines meant that it was no longer enough simply to keep an ear on shipping lanes or convoys: now vast areas of ocean, covering millions of cubic kilometres, had to be monitored. To fill them with sound from active sonar systems would be impractical, so passive sonar was used instead. These systems simply listen for the noises of vessels, rather than bouncing sounds off them. Active sonar is still used for shorter-range detection and monitoring.

During the 1970s, further work and collation of existing knowledge of underwater sound allowed the marine soundscape to be mapped and—in broad terms—understood. It was found that at the lowest frequencies, right down to $^1/_7$ Hz or so, micro-seismic activity, and deep ocean currents are the main acoustic contributors. Surface waves and winds give rise to noise over two ranges: wind causes turbulence, which manifests itself as pressure variations with frequencies of a few Hz, while, at frequencies above about 500 Hz up to 30 KHz, the wind-moved water surface causes the oscillation and bursting of air micro-bubbles. In coastal waters, the impact of tidal streams on seabeds—especially rugged ones—can give rise to noise right up to 300 kHz, though the main source of noise at this frequency is usually the noise generated by the thermal motion of molecules.

However, all over the world, the noise of shipping was also audible underwater, with the vast numbers of ships and the great distances over which their sound travelled merging into a background hum peaking at around 100 kHz. Shipping is often the main cause of underwater noise in the region of 10 Hz–100 Hz.

More distinctly overlaid on all this noise are the sounds of life, in a jungle-like cacophony of frequencies and levels. There was great excitement for a while over the discovery that certain species, such as the croaker, made such distinctive sounds (like a woodpecker in the croaker's case[5]) that a new kind of acoustic fishing might be possible—but, unfortunately, none that was audibly identifiable turned out to be reasonably edible too.

Many fish species communicate by sound, especially when they spawn, but their hearing abilities vary widely, from salmon, which have very poor hearing indeed, to herring, which are highly sensitive to underwater noise.

It was also found that the noises made by living things are used for a whole variety of things, from defence to communication and from mapping to tracking. Toothed marine mammals (*Odontocetes*), especially sperm whales, killers, dolphins, and porpoises, are the most sophisticated users of sound. Whales can make communication sounds that travel for hundreds of kilometres. The frequency range is similar to our own, from 12 Hz to a few kHz (human speech is in the range 50 Hz–10 kHz).

Like bats, *Odontocetes* also make series' of rapid, high-pitched clicks, usually in the 50–200 kHz region, in short bursts of up to ten per second. These bursts allow the animals to sense the distances and directions of nearby objects, and they can even recognize specific types of fish by this means. This shows that these natural sonar systems are more than simple echo sounders; they provide a truly different way of 'seeing'. The wavelengths used by dolphins mean that the sound waves pass through most body tissues of submerged animals and human divers too; what the dolphin 'sees' are articulated conglomerations of skeletons, teeth, and gas-filled cavities like the swim bladders of fish and the lungs of humans. What they make of such views is none too clear. It is, in fact, really impossible to understand quite how important this sense is to these animals, though a hint of its value might be gathered from the many blind people who are able to 'feel' nearby objects by making clicking sounds—allowing them even to take part in games of basketball or to go mountain biking—or to use the sounds of a shower of rain as a way of gauging the spaces, shapes, and surfaces that surround them.

The skills of dolphins in this area seem greater than their ears alone could provide. It is thought that their teeth and jaws may take part in the hearing process too: bottlenose dolphins in particular, which often live in very murky environments where underwater vision is of little benefit, have teeth that seem to act as an acoustic focusing array: they are regularly spaced, similarly shaped, all angled in the same way, and they have heights that depend on their positions in the jaw. It seems plausible that a sound which travels directly along the line of teeth would generate a pattern of constructive interference and so be significantly amplified, allowing the dolphin to home in on such sounds by moving their heads.

While for blind people the use of sound fields as a replacement for visual ones is an amazing adaption, for bats and some marine mammals it is part of life. (Dogs, too, inhabit a very much richer sound world than we do—long-eared varieties in particular make great use of the changing acoustic field as their ears move, to work out directions of sound sources. In fact, despite the fact that in some ways dogs have better vision than humans, they adapt extremely well to the loss of it—so much so that their owners may be unaware for weeks that they can no longer see).

All this means that the impact of unknown noises on marine animals can be very significant. Even if it is not very loud, the addition of background noise to their environments is like our world being gradually filled with a glowing mist—permanently.

Noise on the offensive

In the 1980s, music about the dangers of noise even made it to the charts (albeit not very far up them): Kate Bush's *Experiment IV* told of the devastating accidental results of the test of a acoustic gun, an idea surely suggested by widespread rumours about US military research into acoustic weapons. According to these rumours, the weapons could do just about anything from causing diarrhoea (there was even mention of a 'brown note') to insanity, bone jellification, and apparently anything else anyone could think of. Following the last-minute slow-down of Playschool, the 1988 favourite for the Cheltenham Gold Cup, it was even rumoured that the unfortunate horse had been nobbled by an acoustic gun.

And noise certainly was being used aggressively. In Panama, General Manuel Antonio Noriega Moreno had been a focus of US attention since 1983, when he had seized control of the country. In 1989, US troops invaded Panama in an attempt to oust him. Noriega fled, taking refuge in Panama's Catholic 'Nuncio', which was effectively an embassy of the Vatican. This put the USA in a tricky position. Any military action directed against the Nuncio would not only have been illegal under international law, it would also have been seen as an attack on the Vatican. When the Nuncio staff encouraged Noriega to leave, he threatened to trigger a guerrilla uprising.

To avoid a stalemate, the USA attacked the Nuncio with noise.[6] The fourth Psychological Operations (PSYOP) Group deployed several military vehicles equipped with arrays of loudspeakers, and used them to relay music from the local Armed Forces radio station, which was under American military control. The choice of music (from the American point of view) or noise (from Noriega's) was that of American soldiers themselves, and included such tracks as 'No Particular Place to Go', 'Nowhere to Run', and Rick Astley's (apparently) ever popular 'Never Gonna Give You Up'.[7]

This particular noise attack was not to last for long, however. It was claimed that the noise was preventing reporters using directional microphones to overhear discussions, and the Pope complained about the impact of the sound on the embassy staff. Noriega finally surrendered in January 1990 following protests by Panamanians.

A longer-lasting noise barrage was used against the Branch Davidians at Waco. Selected sounds included rabbits being slaughtered, Nancy Sinatra singing 'These Boots Are Made for Walking', jet engines, and, strangely, the chanting of Tibetan monks—all of which were played, loudly and with little intermission, for fifty-one days.

A more peaceable application of noise was the destruction of kidney stones, previously treatable only by unpleasant surgery or even more unpleasant waiting until the stones passed through the urethra (said to be one of the most painful of all experiences). Carried out experimentally in the early 1980s in Germany by Dornier Medizintechnik GmbH, this technique, called lithotripsy, became common following the introduction of the HM-3 lithotripter in 1983. Up to 120 shocks per minute can be delivered fairly accurately, and the shocks both tear the stones apart directly and wear them away rapidly through the formation of bubbles in a process called cavitation. The tiny fragments can then pass from the body with the urine. It is also possible to treat gallstones in this way, but, as they are less brittle, it is not so effective.

Perhaps the oddest power ascribed to sound at this time was that of initiating nuclear fusion. The idea of 'cold' fusion was simple: palladium was immersed in water and subjected to ultrasound. Supposedly, this would initiate a fusion reaction of hydrogen from the water on the metal surfaces, being detectable as a temperature rise and as a flow of neutrons. Given that ultrasound can generate pressures and temperatures like those deep inside the Sun, albeit within tiny volumes, the idea is not as strange as it may appear.

Unfortunately, one of the significant problems with ultrasound is that it is very hard to measure its power, particularly as the area over which the ultrasound is delivered is very tricky to control, being easily reflected or otherwise changed by features of the medium. In fact, in the early days of lithotripsy the power of some lithotripters was assessed simply by aiming them at a plastic cup and seeing how fast it melted. So, measuring the heat accurately enough to tease out a fusion-produced component is extremely challenging. In fact, it was the lack of neutrons that showed that fusion had not been initiated in this case.

Interest in low-frequency noise also grew during the 1990s. For long it had not been taken seriously by the acoustics community because of the great difficulty in verifying the existence of such sounds, but more serious efforts did pay off, and it was found that in fact they were often measurable by conventional sound level meters, but varied in strength over short distances—so a low-frequency hum around the head of a bed that is loud enough to wake a sleeper might be completely inaudible and immeasurable a metre away. So, many people who had been written off as having good imaginations—or even as being malicious timewasters—were suddenly regarded in a more sympathetic light. The problems of such sounds are, however, not easy to resolve: for one thing, their sources are very difficult to determine, because low frequencies are highly non-directional anyway, and because they are transmitted by structures they can travel in very non-linear paths. And, since they *are* structure-borne, keeping them out is very difficult too without major structural modifications. Finally, even if the source can be tracked down, it often turns out to be an underground train system or something similarly tricky to modify without vast expense. Thankfully though, such problems are more frequently caused by local effects, such as faulty air-conditioning units.

An infrasonic mystery from the 1990s is the Bloop—a very powerful, very low-frequency (from a few Hz to around 50 Hz) underwater sound detected by the US National Oceanic and Atmospheric Administration several times in 1997 and occasionally still recorded to this day. It apparently originates far out in the Pacific Ocean, somewhere west of the southern part of South America. It lasts about a minute and is loud enough to be detected by sensors located over 5,000 kilometres apart. In form it is vaguely similar to the sounds made by marine animals, but is many times more powerful than the call of the blue whale.

It is not only in the dark depths of the sea that mysterious sounds can be heard. One of these, famous since the late 1990s, is the Largs Hum (Largs is on Scotland's West Coast, overlooking the Firth of Clyde). Unlike unknown underwater sounds, land-based ones like this are generally continuous. The Hum has proved impossible to localize, and it varies in strength oddly, often becoming much louder in cars and buildings, presumably because of resonance effects and standing waves. Even more strangely, the Hum affects TV and radio too, which suggests that its source is electromagnetic.

The Hum seems to be just around 20 Hz, it cannot be heard by everyone, and it is easier to feel than to hear. Like other low-frequency sounds, it causes a feeling of pressure round the eardrums and chest, adding to its unsettling effects.

Similar hums have been reported in many other places, including Bristol and Taso, New Mexico, where a hum began in 1991 and has continued to this day. A survey of Tasoians in 1994 showed that 11 per cent of them could hear it. The Bristol Hum may be the most persistent of them all. It has lasted from the 1970s until the present, and even drove a local resident to suicide in 1996. Such hums are mostly infrasonic, or have a large infrasound component, and beats are often present too—suggesting that multiple sources may be involved. A pronounced hum is also notorious in the Netherlands, though its source is no mystery: low-frequency sounds generated by the constant stream of ships passing close to the coast travels easily through the saturated soil, disturbing people living up to 20 kilometres inland.

The Netherlands hum is well understood, thanks to the now highly developed science of underwater acoustics. Continued funding for the area stems from its central importance to underwater military activities. For example, with the ending of the cold war in the late 1980s, it was not just relations between Russia and the West that were warming up: submarines, designed to be deployed in the cold waters of the North Atlantic, were now gradually being deployed in the Gulf, especially in 1990–1 when the first Gulf War took place. Since the first experiments in Germany in the 1930s, efforts had continued to develop sound-absorbing materials that could be used to cover a submarine, thus rendering it invisible to sonar systems. By the 1980s these stealth coatings were highly sophisticated,[8] but unfortunately the materials are rubbery ones, and their physical and hence acoustic characteristics are highly sensitive to the temperature and pressure of the water that surrounds them. So coatings that are almost perfectly non-reflective in the deep cold waters of the North Atlantic became rigid reflectors in the shallower, warmer waters of the Gulf. Rapid research was needed to develop replacements.

Noise worldwide

Research into the effects of noise on people continued, with special emphasis on the impact on children. The influential RANCH project[9] investigated the effects of road traffic and aircraft noise on school performance, annoyance, and blood pressure in 9- and 10-year-olds who lived near major airports in the Netherlands,

Spain, and the UK. Though chronic exposure to aircraft and road traffic noise was not found to impact mental or physical health, it was found that aircraft noise exposure interfered with both reading comprehension and (recognition) memory, as well as causing annoyance. Reading age in children exposed to high levels of aircraft noise was delayed by up to 2 months. It was not possible to ascertain whether these effects were temporary or permanent.[10]

Previous studies had drawn similar conclusions. In New York in 1975 Arlien Bronzaft, a psychologist, conducted a study to track the reading skills of a group of students who were exposed to constant noise pollution against a group of students who had a quieter academic environment. The result of the study revealed that those students who experienced noise pollution had a noticeable reading delay, whereas the students in the quieter environments did not. However, the RANCH project was one of the first to establish a dose–effect relationship between noise and performance—that is, it showed that the effects increase as the noise does.

The 1990s was a significant decade for the fight against noise, with increased efforts to take international action, especially across Europe. The World Health Organization Regional Office for Europe set up a task force in 1992, reporting in 1995 and recommending that people should have the right to decide the 'quality of the acoustic climate'—their soundscapes, in other words. This publication served as the basis for globally applicable Guidelines for Community Noise. The next year, the World Health Organization declared that noise was a significant threat to human health.

In many ways, the European Union (EU) is the star of the show as far as noise is concerned and has replaced the USA as the world leader in noise control. EU policies are not only forcing national governments to take the noise issue seriously, but also providing a much-needed framework describing how to tackle it. The fact that such agreement has been reached is an impressive achievement, given the cultural differences involved, which, as discussed above, directly impact on the way in which noise is regarded and even measured.

One might hope that the availability of noise-measuring instruments and the carrying-out of noise surveys for almost a century would have led to an objective history of noise levels, describing how, where, and when they have changed worldwide and country by country over the last few decades. But there is no such thing. Differences in the measurements made or the questions asked usually make proper comparison impossible. Data collected in preparation for the 1996 EU Report 'Future Noise Policy' show just how difficult it is to be sure even whether noise has increased overall.[11]

The report's authors estimated that one in five of the EU population was exposed to excessive levels of noise, defined as being above 55 dB Lden.[12] In large urban areas this proportion rose to over 50 per cent; 15 per cent of city-dwellers were exposed to noise over 65 dB Lden. Statistics collected since then show no clear changes, and what information is available from earlier decades also reveals few trends. Noise levels related to traffic noise were stable over most of the 1980s, even though action over 'black spots' where noise had been over 70 dB had often been successful. In many countries, though no more people were being exposed to noise above 65 dBA, increasing volumes of traffic meant that more of them were experiencing noise in the 55–65 dBA region. In urban areas in general there was no increase in traffic noise peaks in the 1980–95 period, but the time over which the roads were noisy was being extended, with a particular increase in night-time noise.

Overall, EU surveys revealed that up to 170 million of its citizens were living in areas with sufficient noise levels to cause serious annoyance during daytime, one in four of its population suffered reduced quality of life, and between 5 per cent and 15 per cent of the EU population experienced serious sleep disturbance. And it was not simply annoyance and loss of sleep that was the problem: in 1996, the Royal National Institute for Deaf People estimated that more than 6.5 per cent of the population between the ages of 16 and 60 were mildly to profoundly deaf—with most of this hearing loss being ascribed to noise.

The one area where really significant progress has been made since the 1970s is aircraft noise, thanks mainly to stricter noise certification standards. For instance, the number of people around Heathrow exposed to noise levels above 60 dBA more than halved between 1975 and 1989, even though there was a significant growth in traffic over the same period. Large decreases have also been experienced at Copenhagen and at Schipol airport in Amsterdam. Moreover, the noise arising from aircraft was only one-ninth that of 1970s planes of similar size.

On the railways, noise emissions from individual trains have also fallen, largely thanks to the changeover from diesel to electrically powered passenger trains, the gradual introduction of welded rail to replace jointed rail, and the greater use of disc braked rolling stock. Despite this, railways do have their problems: in recent years the use of warning sounds on approaches to level crossings has caused problems across Europe. In efforts to reduce noise impact in general, the hooters used have been made more directional—which is fine, except when there is a curve just before a crossing. When there is, the unfortunate people living straight ahead of the train's path have to be blasted with extremely loud sound to ensure that those just around the corner at the crossing can hear it clearly.

Underground trains are even more problematic in some ways: metro systems are horribly noisy in many cities, a problem that is surprisingly difficult to solve, considering that the whole environment is a carefully planned and constructed one. Nevertheless, many highly expensive projects in many cities (in particular Paris and Berlin) have failed to make significant headway against many of the noises heard there.

On the roads, regulatory and technological advances have struggled to keep pace with the increasing number of vehicles. Travel related to tourism and leisure activities in particular has increased significantly. So, while the noise from individual cars has been reduced by 85 per cent since 1970 and the noise from lorries by 90 per cent, road noise is almost as much of a problem as ever. Lorry noise is still the dominant source in New York City, and noise is still the most common subject of complaints to its City Hall.

In the UK, the 1989 Noise at Work Regulations (which implemented an EU directive) introduced an eminently practical system of 'action levels', which nicely involved both employer and employee in the process of noise projection. If the average daily noise level exceeded 85 dBA, then the employer had to provide hearing protection if the employees requested it. At levels over 90 dBA, the onus shifted to the employer, who had to provide not only protection on their own initiative, but also warning signs and training. In addition to the overall total level of acceptable exposure, limits were set according to the maximum pressures reached by single loud noises (at 140 dB). At the time, the first action of 85 dBA was thought to be present at more than 80,000 workplaces. In 1994, The UK Health and Safety Executive estimated that around 845,000 workers were exposed to noise above this level.[13] In 2003, the UK Physical Agents (Noise) Directive reduced these levels of acceptable noise, typically by 5 dB.

Meanwhile in France the first law covering most aspects of noise pollution was passed in 1992. Still in force in 2012, it covers town-planning, transport systems, and noise protection near airports, mandates proper noise measurement programmes, and imposes significant penalties on offenders.

16

NOISE NOW

Environmental noise has become a broad and deep subject, with vast numbers of measurements, regulations, plans, complaints, and problems. Many of the issues are the same as those that have dogged us for years, but there are a few that have a peculiarly contemporary relevance.

In the twenty-first century, as in previous ones, it has been new noise sources that have made the headlines and caused annoyance, from night-time rickshaw music in London to the sound of vuvuzelas in South Africa during the 2010 World Cup. It was not only the novelty of the latter that caused unease in the hearts of the noise-wary. So intense is the sound of a vuvuzela that one just behind you will insonify your head with something like 125 dB, and probably make you jump too. Add this to the other loud noises of cheering, shouting, and jeering so prevalent at World Cup matches and the cumulative effect can be very significant indeed—so much so that the then Royal National Institute for Deaf People (RNID) warned those attending the matches to defend themselves by wearing earplugs.

Another newish noise is the whir of wind turbines punctuated by the 'slap' as the turbine blades pass the tower. This too is not just a new noise; it is also an unusual one in several troubling respects, and is likely to become more and more widely heard in future, not least because of the retreat from nuclear power following the

Fukushima nuclear power station crisis. The sound radiated by a wind turbine is quite low powered, but it is especially intrusive in its often rural setting. Since the noise source is overhead, conventional barriers are pointless, and since wind farms are sited on exposed locations, there are unlikely to be any natural barriers nearby. Other problems are the unpredictability of the wind, and hence the noise, together with the unfortunate fact that winds blow more by night than by day.

In many countries, wind farms are the focus of disagreements not just between local and central governments or between environmentalists and others. They also set environmentalists at each other's throats, forcing them to weigh their own arguments against each other. Does visual amenity matter more than quiet? What is more important: nesting birds or carbon footprint? Unfortunately, the process of decision-making that results is not very edifying. Whoever shouts the loudest often wins, and final decisions are generally made at a local level, which means that no national policy has a chance to evolve. Of course, the answer is not that we need to decide which issue is most important but to come up with a solution that pays due and balanced regard to all. Without such compromises, only conflict, inefficiency, and alienation can result.

There can now be hardly anyone who doubts the impact of noise on humans, but there is still one major research gap: the difficulty of applying dose–effect relationships to realistic situations. While there are many well-proven relationships for individual noise types, they have mostly been developed in controlled or specially selected environments, where there is only one noise variable to consider. In real situations, not only do a vast range of other parameters affect the results; there are frequently several noise sources too. The interaction that results is so complicated that it is all but impossible to predict the effects quantitatively. Without such knowledge, financial and other quantitative impacts of noise and noise reduction cannot properly be determined, and that means that cost–benefit analyses cannot be done. So, since local and national authorities do not know just how much less of a bang they are getting for their bucks, they are discouraged from committing substantial funds to noise reduction projects.

Not all noise research and development is intended to reduce noise impact, however. The *Mosquito* (invented by Harold Stapleton in 2005, first sold in 2006 through his company Compound Security Solutions, and called in France, rather disrespectfully, the *Beethoven*) emits a sound with a frequency around 17.5 kHz that only the young can hear—and it is not a nice one. The idea was to discourage teenagers from hanging around outside shops. First trialled in 2005, the *Mosquito* is still in occasional use today, though it has caused great contention: while it has been used as a condition for granting licences to UK off licences, some councils, including Westminster and Sheffield, are opposed to its use, mainly because it cannot discriminate between babies and hoodies. In fact, the Council of Europe voted to ban the device in June 2011, but the ban is not binding unless the European Parliament vote to enforce it.

Cunningly, the victims of the *Mosquito* have used the idea to fight back, using high frequencies as 'teen buzz' mobile phone ringtones—so that their ageing (over 30, that is) teachers cannot hear their phones ring in class. And, of course, it is not just in schools that a battle for mobile phone rights has been joined. Debates still continue about whether it is acceptable to use a mobile in a restaurant, answer a call when out with friends, or conduct teleconferences on the train. But zoning has come to the rescue here: in many countries trains have a special 'quiet carriage', designated as a mobile-free area (and, on commuter lines in Boston, mobile phone use is forbidden during rush hours).

The future of public mobile phone use is hard to predict—on the one hand, it may be that local signal blocking systems are installed in more public places, or, on the other, people may accept mobile ringtones and conversations as just another element of background noise.[1] Perceptions of the expected level of background noise are certainly changing in some locations, such as libraries, and young people today are able to cope with multiple simultaneous inputs—computer, CD, iPod, TV, mobile, parental lecturing—in a way that few people of the previous generation would ever have needed to.

While some might say that the *Mosquito* is a genuine nuisance and an infringement of minority rights, others might regard it simply as another example of that extraordinarily well-publicized twenty-first-century phenomenon, 'health and safety gone mad'. This latter group might well be similarly unconvinced by the 2006 decision that Scottish soldiers learning to play the bagpipes must wear earplugs and limit practice sessions to twenty-four minutes a day outdoors or fifteen minutes indoors. This followed a study by the Army Medical Directorate showing that the sound could reach 111 dB (A-weighted, I'm hoping) outside, which is slightly louder than a pneumatic drill, and 116 dB inside, which is as loud as a chainsaw.

Unhearable, uncanny

At the other end of the spectrum, the range of known infrasonic phenomena is growing larger and more strange. It might even hold the explanation for at least some ghostly phenomena. In 1998, Vic Tandy, lecturer in the School of International Studies and Law at Coventry University, and Dr Tony Lawrence of the Psychology Department wrote a paper called 'The Ghost in the Machine' for the *Journal of the Society for Psychical Research.*[2] Appropriately enough for two academics, their account was not of a haunted house but of a haunted laboratory. Accounts of late-night visitations to scientists there spoke of blurry grey figures, and there were frequent reports too of feelings of dread and depression. A clue to a non-supernatural explanation came when Tandy was repairing his foil (he was a keen amateur fencer). He found that it vibrated alarmingly from time to time—but only in particular spots in the laboratory. After several fruitless attempts to work out the cause, Tandy tried switching off an extractor fan. The results were immediate: 'It was as if a huge weight was lifted,' he said. When the infrasound was measured, it was found to be at just the right frequency to make eyeballs vibrate and so perhaps to generate visual illusions.

The emotional effects of infrasound were studied in more detail by a team including staff from the National Physical Laboratory, Liverpool Hope University, and a number of engineers, performers, and a psychologist. The team installed an infrasound generator—a pipe 4 metres long and 1 metre wide with a loudspeaker at one end—in the Purcell Rooms in London in May 2003. Two concerts of the music of Philip Glass and other contemporary composers were performed, during one of which the infrasound pipe was activated. Questionnaire responses from the audience suggested that, with infrasound present, emotional reactions to the music were stronger—whether they loved it or hated it, audience members loved or hated it more. Some even spoke of shudders down their spines, coldness, anxiety, and sorrow. This effect is consistent with the findings of Tandy and Lawrence in 1998[3]: if someone in a supposedly haunted house is in a nervous frame of mind already, then their fear state could be enhanced by infrasound sources, which might, for example, be generated by the wind blowing across chimneys. It is perhaps no wonder, then, that a storm is often the background to the scarier moments in films and stories about hauntings. And perhaps, too, Robert Clavering was more on the mark than he knew when, in 1779, he referred to the gloom-inducing effect of winds howling down chimneys. Whether all this is true or not, the world of infrasound is one that still has secrets to reveal.

A unified approach?

Partly in view of the differing national conditions and concerns cited previously and partly because of the general paucity of applicable dose–effect relationship data, attempts to reach international agreement on acceptable noise levels are still floundering. The most promising avenues at the moment are those that concentrate, not on the topic of annoyance, which is so hard to pin down, but on the rarer but better-defined physiological effects. While it is often not possible to ascribe a physical effect definitively and solely to a noise cause, one area where a clear pattern has emerged is cardiovascular ailment: in 1999 the World Health Organization (WHO) defined the noise threshold for the onset of such problems to be a long-term night-time level of 50 dBA. More tentatively, it was concluded that the threshold for sleep disturbance is 42 dBA, and for general annoyance, 35 dBA.

Classifying noise as a pollutant has many advantages. It adds to the seriousness with which it is treated and encourages people to try to apply solutions that have been used to tackle other pollutants. It also highlights the fact that noise cannot be tackled without considering the impacts on other pollutants, and vice versa. This was recognized by some as long ago as the 1920s, when it was noted that the popularity of opening windows to increase the flow of fresh air, while a positive act in itself, increased the negative impact of noise. Similarly, mitigating noise without attending to the impact of that mitigation on other concerns can be at best inefficient and at worst counterproductive. No one wants to look at ugly concrete noise barriers running along nearby motorways. But, with sufficiently strategic planning and the cooperation of different authorities, local inhabitants might instead be able to enjoy both reduced noise and the view of a dense strip of vegetation: 100 metres of natural forest attenuates noise by about 20 dB. In addition to this objective reduction, an additional 5 dB's worth of subjective improvement is indicated by a 1997 study.[4] A Swiss/German study supports this by showing that the number of people who are annoyed by traffic noise is significantly smaller—all other factors being equal—in more attractive streets.[5]

However, it is only since the late twentieth century that a holistic approach to dealing with noise and other pollutants has been acted on at a national or international level, confirmed as law by the Integrated Pollution Prevention and Control (IPPC) Directive, which requires that all types of pollutant should be dealt with in a joined-up way.[6] This type of thinking is still new, and local authorities in most countries usually still have different departments dealing with different pollutants, with relatively little coordination between them.

The Netherlands is an exception here and is, in terms of noise research and control, one of the world's most active countries. This is due in part to its high population density (seventeen million in 33,883 square kilometres) and the concomitant noise problems that causes, with about one in three people bothered by neighbour noise and 75 per cent of dwellings exposed to noise over 50 dBA. In 2003, 26 per cent of the entire Dutch population reported that they were 'highly disturbed' by noise while sleeping.[7] Unfortunately, noise travels very readily from place to place too, because of the water-filled subsoil and the lack of high ground to act as a noise barrier.

Shocks of sound

As science continues to study the impact of noise on people, it becomes ever clearer just how complex and unexpected are the reactions involved, and none more so than to acoustic shocks. An acoustic shock is a sudden, unexpected sound heard over a telecommunications system—often caused by a burst of static, a click of disconnection, or an abrupt electronic tone. It has been a particular concern to the burgeoning industry of call centres, not only because their workers constantly use phone lines, but also because they are connected to them by headphones, which cannot be removed quickly if they transmit an unpleasant sound.

The consequences of acoustic shock can be profound, from tinnitus to stress, insomnia, a range of hearing disorders, and a real phobia about continuing with the job because of the concern about the shocks themselves.

What makes acoustic shock such a challenging noise problem is that the effects are produced even by quite quiet sounds—ones with sound pressure levels much lower than the statutory maxima. They are lower, in fact, than some speech sounds, which means that simply quietening the line would seriously interfere with conversation. An additional complicating factor is that acoustic shock seems to affect some people much more than others.

Though this has not received a great deal of publicity, the effects of acoustic shock have led to numerous claims against employers. Though none has yet come to court, out-of-court settlements have exceeded £2 million in the UK and £10 million worldwide since the 1990s.

In 2011 the WHO estimated that about one million healthy life years[8] are lost every year because of traffic noise in western Europe, through cardiovascular disease, cognitive impairment, and the stress-related illnesses resulting from sleep disturbance, tinnitus, and annoyance. To take a specific example, coronary heart disease killed 101,000 in the UK in 2006, and, according to the WHO's analysis, 3,030 of those deaths were caused by chronic exposure to traffic noise. Overall, the noisiest EU country is Finland, and the least noisy is Italy.[9] Significantly, insufficient data are available from other countries for the WHO to draw such conclusions outside western Europe.

The chief source of noise problems in terms of numbers affected is, by far, road transport. The fraction affected, however, varies a great deal from country to country—only about 4 per cent of the population are exposed to daytime road traffic noise in excess of 65 dB in the Netherlands and 10 per cent in the UK. This is 10 per cent too many, of course, but UK citizens might spare a thought for Spaniards, of whom a whopping 57 per cent are so exposed.[10] Overall, about half of the EU population who are living in urban areas with more than a quarter of a million inhabitants have to cope with noise levels above 55 dB Lden, from road noise alone. Children in London schools are regularly exposed to noise that exceeds WHO Guidelines, and it is now quite clear that this can adversely affect their performance.

The most recent large-scale noise survey in the UK was the 2008 National Noise Survey, conducted by Ipsos MORI on behalf of Environmental Protection UK, and involving around 2,000 adults. Of those people surveyed, 25 per cent were bothered by noise from neighbours, 13 per cent had been woken up by noise caused by neighbours in the previous year, and 1 per cent of people had moved house because of noisy neighbours. Perhaps not surprisingly, while 4 per cent of the respondents said they had argued with a neighbour about the noise they were making, only 2 per cent said that a neighbour had argued with them about it.

In terms of which noises were most annoying, as one might expect, cars and motorbikes were most frequently referred to (18 per cent), followed by—in joint next place at 12 per cent—'car/burglar alarms', 'fireworks', and 'children'.

Such evidence as there is of recent changes in noise level indicates that they are slight. The UK Building Research Establishment carried out a national study of environmental noise levels for the Department of the Environment in 1990 and again for the Department for Environment, Food and Rural Affairs and the Devolved Administrations in 2000 (though only in England and Wales). Measurements of the noise over 24 hours outside 1,000 dwellings were made. The studies showed that changes in noise levels over the decade were small, and, while noise had fallen slightly in some places, it had increased slightly in others—which is consistent with the finding that people interviewed felt that there has been no change to speak of, other than regarding neighbourhood noise. This lack of improvement is of concern, since both surveys showed that the majority of the population were exposed to noise levels above those recommended in the WHO Guidelines for Community Noise.

It is harder to draw any conclusions about noise change in the USA, because of the dearth of recent data. In 1980, it was estimated that 16.2 million US citizens were exposed to average noise levels of 85 dBA or over. In 1990, the number exposed to levels over 55 dBA was 138 million, with 25.4 million above 65 dBA, and 1.4 million above 75 dB.

This lack of recent change may indicate that the many laws, regulations, and technological solutions developed to deal with noise have already been applied as far as practicable, and that further progress can be made only through larger-scale, longer-term strategic initiatives.

By far the most important such initiative of the early twenty-first century is an EC directive, 2002/49/EC, 'Assessment and Management of Environmental Noise', usually known as the European Noise Directive or the END. The END requires all member states to construct noise maps of all major conurbations and transportation routes. (Fig. 40 shows an example of a noise map of an urban area.) These maps do not attempt to capture the actual noise levels at particular times, but rather to determine the overall impact of (mostly) transport-related noise sources, averaged over different weather conditions. The maps are defined by calculations, based themselves on such inputs as transport flow data. The noise maps are then supposed to be used to develop action plans to reduce high levels of noise, especially in noise 'hot spots'. The END does not set any limit values, nor does it prescribe the measures to bc used to develop the action plans, which remain at the discretion of the competent authorities. One of its key legacies was to establish, after decades of previous attempts to do so, a unified approach to noise assessment across most of the continent.[11] Partly as a result of this, and partly because of the new-found seriousness with which noise was being taken, many countries have passed new laws that attempt to curb noise.

As well as reducing high-level noise, the END mandates member states to identify and maintain quiet areas in cities and in open country. The decision as to what constitutes a 'quiet area' is left to the national authorities (who often, as in the UK, then pass responsibility to local authorities). This is no doubt a very sensible way to approach the problem: defining a noisy area can be difficult

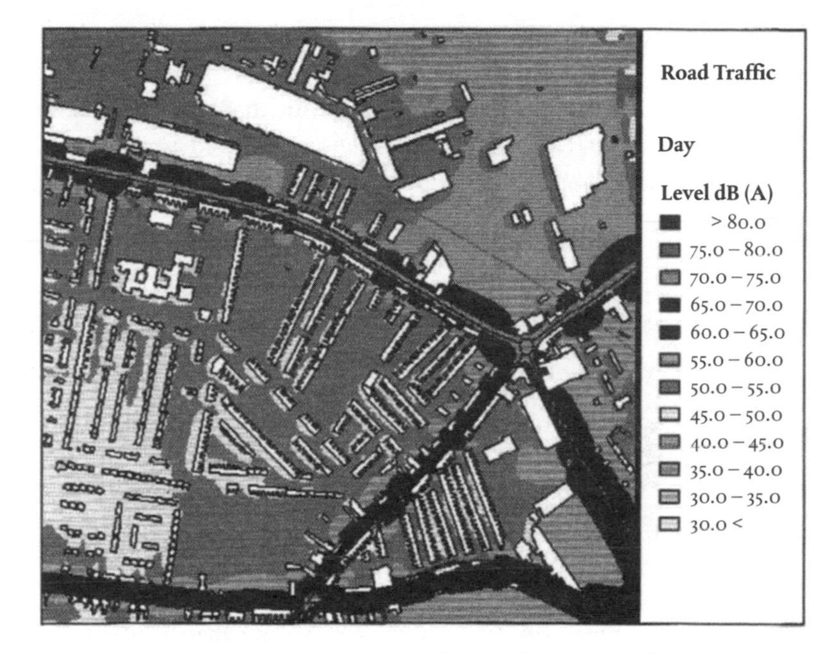

FIGURE 40. A noise map, based on traffic noise predictions.
D. Knauss, 'Noise Mapping and Annoyance', *Noise* Health, 4 (2002), 7–11. Medknow Publications and Media Pvt Ltd.

enough—defining a quiet one in a useful way is far harder, largely because a place where there is little noise is of no benefit to anyone unless it is also reasonably attractive, accessible, convenient, safe, healthy, and supplied with necessary amenities. And, of course, what is attractive to the inhabitants of one region may be most unattractive to those of another—nor do the inhabitants of different countries, with different preferences regarding transport, all regard accessibility in the same way. And so on.

In addition to appointing a team of international noise experts (the Noise Expert Network) to help with detailed noise policy development, the Commission also initiated a series of international research projects and toughened up the enforcement of noise emission standards, especially those relating to vehicles.

Noise Action Plans were to have been drawn up by July 2008, setting out measures to reduce noise, and submitted to the European Commission by December 2008. Most member states fell behind in the process but have now published them.

Like so many other highly ambitious international projects, the END and its implementation have come in for criticism, and are hampered by some significant limitations and subject to many delays. Nevertheless, the project represents by far the largest, best organized, and—arguably—most promising attempt to tackle noise problems the world has yet seen.

As might be anticipated given the numerous failed attempts to reach international agreements on approaches to noise measurement, one of the most challenging aspects of the END projects has been to define approaches that are common across different member states. The lack of consistency means that some maps made by different states that impinge on each other can differ by as much as 10 dB in their predictions for the same point. One can well imagine too that work on the project is likely to fall victim to more pressing financial problems in many member states over the next few years.

Lack of funding is, however, not the only problem. While the EU directives have been obeyed—albeit rather late—some might be forgiven for getting the impression that they are regarded by government bodies and some local authorities as an unnecessary bit of eurocracy, instead of a ground-breaking achievement and the most impressive concerted effort to deal with noise there has ever been.

Certainly, many UK organizations who should know better seem unpersuaded of the importance of noise pollution: in 2006, for instance, the Department for Communities and Local Government commissioned research on green spaces—but with no mention of noise. The 2007 Royal Commission on Environmental Pollution, the 2008 Mayor of London's consultation on best practice on the development of strategies for open spaces, and Natural England's consultation on a policy for landscapes all also excluded noise from their terms of reference strategies.

A change of mood

The studies carried out around the end of the twentieth century in the UK at last provided good benchmarks for measuring real changes in noise levels. All that was necessary was to wait and do some more similar ones. Or so one might hope. In 2007, the UK government carried out a survey whose results made the question 'Is the level of noise increasing?' rather less pertinent. The results of the Department of Trade and Industry's Attitudes to Noise from Aviation Sources in England (ANASE) survey were published, after long delays, in November 2007 (and then only after questions had been asked in parliament about the delay). At this point the government made it clear that the results were not really that interesting; in fact, they were hardly worth looking at; it wasn't sure why it had bothered to get the survey done in the first case, and it certainly wouldn't be worth taking the results into account in deciding future policy. So one could tell something significant had turned up. In fact, what the study showed was that people were substantially more sensitive to noise than they had previously been. Since the only point of quietening airports and other such noise sources is to reduce their effect on people, the clear implication is that such sources need to be quietened still further than planned, to deliver the required reductions in disturbance. This was reminiscent of the finding of the 1967 survey of those living near Heathrow that 'attitudes had hardened' since 1961. Unfortunately, there are no comparable data from other countries, so it is hard to judge whether this is a peculiarly English change.

One can only speculate about the reason for such a change in attitude. One possibility is that this is a sign of the latest shift in society as to who holds the whip-hand, and is therefore empowered to complain about other noise-makers. Certainly there is a much reduced level of public confidence in science and scientists, reflected, for instance, in the rejection of GM foods and the enthusiasm for organically grown ones, in spite of the scientific evidence that such reactions are baseless. Whether this is the case or not, there is certainly more public interest in noise and its problems than there has been at least since Concorde—or the idea of it—was in the air. It will be interesting to see, therefore, whether there is much interest in the EU-mandated noise maps of European towns and cities and the action plans based on them, once they become better known, more reliable, and more widespread.

In the UK in the early twenty-first century, the way in which noise is tackled is complex. The general approach is still, after many centuries, based on the laws of public and private nuisance. Local authorities have the statutory powers to investigate, prevent, and remedy noise. There are some criminal offences based on noise, including night-time noise, and the police have the power of confiscation in this case. There are regulations regarding aircraft noise, noise at work, noisy parties, public entertainment, burglar alarms construction sites, industrial sites, and noise on public streets.

However, in Francis McManus' 'Noise Law in the United Kingdom—A Very British Solution?' it is made clear that national strategies regarding noise are weak and vague, that enforcement varies widely from one local authority to another, and that local authorities are not strongly committed to combating neighbourhood noise.[12]

In the USA, the closure of the Office of Noise Abatement and Control in 1982 is still being widely felt: the lack of a central body to organize, encourage, and develop actions against noise has led to stagnation and uneven enforcement. While many city ordinances now prohibit sound above a particular set level from trespassing over property lines (with different levels at night and during the day), in many of them complaints are often not followed up and when they are all that usually happens is that warnings are issued. Any real actions, in the sense of court proceedings, are very rare, apparently because of their cost—and it is just possible that the knowledge of this rarity might reduce the effectiveness of the warnings somewhat.

After major progress in earlier decades with respect to aircraft noise, that area too has fallen into disrepair with the demise of the Office of Noise Abatement and Control, which had the responsibility for legislating against such noise sources as planes, trains, and lorries. In 1981, it had not yet finished the job: in particular, it had not got round to putting in place standards for aircraft noise, and the state authorities were not empowered to do anything either. The only organization that is now permitted to set such standards is the Federal Aviation Administration (FAA), which is rather keen on planes and probably is not that bothered if they are a bit noisy. Though often challenged in court when new or extended airports are planned, the FAA is higher in the pecking order than is the Environmental Protection Agency (EPA), and, though it is mandatory to assess noise impact, there is no requirement to make planning decisions on this basis. As in the UK, instead of this the FAA spends considerable sums on sound insulation.

Furthermore, as a result of the power of the automobile in American life, states have no power to set limits for individual vehicles, and the national threshold levels that do exist are only guidelines. On the other hand, noise regulations are firmly in place in the cases of new roadways and other transport installations, since the National Environmental Policy Act and the Noise Control Act have long required that new transport systems must produce less noise impact than those they replace.

The lack of coordination across the USA does not, of course, mean that no one is doing anything about the problem of noise. Many individual states are still world-leading in some respects. Portland, Oregon, for example, is a shining example of good noise management—it has had a very effective Noise Control Office since the late 1970s and a noise code that is among the best in the world.

The emphasis in Europe on the production of noise maps as a starting point for noise control projects has been echoed in recent years in the USA, where noise maps of national parks are being drawn up as part of the Natural Sounds project. The purpose of the maps is to help manage over-flying by air tour craft (a key concern of the National Parks Air Tour Management Act of 2000). The first such map was completed in 2009, along with a detailed plan to maintain the equivalent of Quiet Areas. And there is no doubt that such actions are needed: the Grand Canyon, for example, which until the 1950s was a byword for peace and serenity, was over-flown by 50,000 flights in 1987. In response, the National Parks Overflight Act of the same year required 'substantial restoration of the natural quiet', and Bill Clinton even issued an executive order to the Park Service to make this happen, in 1996. And the result? In 1998, 132,000 flights over-flew the park. Meanwhile, in Yellowstone National Park there are continuing debates over the noisy use of snowmobiles, and the use of the vehicles is now heavily controlled there. Not that everyone is on the same side, of course—in 1990 Gale Norton, soon to become Interior Secretary, even argued for a 'homesteading right to pollute or to make noise', in a bid to support the rights of private property-owners.

Meanwhile in 2007 New York City council revised its existing noise code (set up in 1972—the first in the USA). In addition to tightening up existing restrictions, it also introduced an innovative and controversial element to modern noise regulation: that transient sources such as fireworks, motorcycles, and so on can be cited by a police officer for noise violation simply on the grounds that the sound is 'plainly audible'—that is, even in the absence of any noise measurement. It remains to be seen how well this will work.

Often, the development of effective anti-noise regulations is a slow process, with early versions being relatively narrow, weak, and sometimes not as appropriate as they could be. Madrid, for example, is anecdotally the noisiest city in Europe, and it is probably no coincidence that Spain introduced national noise limits only in 1993.

In France, according to data gathered by the French Environment and Energy Management Agency, the majority of citizens consider noise pollution to be their principal everyday concern.[13] Transport noise accounts for 80 per cent of noise pollution, and 68 per cent of transport noise is from road traffic.

While road transport noise is usually the most significant type in any large city, the type of vehicle varies widely, from lorries in New York to mopeds in Greece. There are 12 for every 100 people there, twice the EU average. An additional problem in Greek cities is that traffic is often redirected down small narrow streets by the police to reduce traffic jams on main thoroughfares, and in many cities the noise made is enhanced by reflection from the high, tightly packed buildings.

Outside Europe and the USA, both the types of noise regarded as polluting and the actions taken against them can be very different. In China, for example, cities are often blighted by the noise of construction as well as of transport: twenty-four-hour building work is now commonplace in Shanghai. In recent years the popularity of domestic stone furniture has led to many noise problems, since it is frequently constructed *in situ*. On the other hand, since 2011 the 113 largest Chinese cities have been required to report the amount of noise made within their precincts on large public hoardings, to install networks of noise-monitoring devices, and to identify major noise sources.[14]

In all cities undergoing rapid growth or development, construction activities are frequently the dominant noise source in many areas. Not only is it almost impossible either to reduce the noise at source or to contain it; construction work also both increases and displaces traffic flow, spreading the noise impact over wider areas. Such noise is also very hard to predict, and the dynamic nature of the projects means that local authorities have little time to react to it. It was largely because of construction work and its impact (along with aircraft and other transport noise) that 70 per cent of Moscow was classified in 2010 as a zone of noise discomfort by the Environmental Health Service. In some of the busiest Moscow streets, noise levels are frequently above 85 dBA—a level that, in the EU, would mandate the wearing of hearing protection.

In Sydney, while construction noise is not a major source of complaint, its products are: as more high-density housing projects are completed and occupied by people leaving more spacious and separate suburban dwellings, neighbourhood noise problems grow: most of the city's 100,000+ noise complaints per year relate to them. Meanwhile, in Melbourne, noise complaints in one area doubled after the introduction of higher-density housing.

In Japan, the soundscape concept has proved popular and seems to be regarded as a more obvious approach to dealing with unwanted sound than in Europe or the USA, perhaps in part because of the long tradition of including sound-making water features in Japanese gardens. The 'One Hundred Soundscapes of Japan: Preserving Our Heritage' project, launched by the Japanese Environmental Protection Agency in 1997, has proved to be an effective way of dealing with at least some noisy areas: the popularity of the project and the publicity (and resultant tourist interest) accorded to the 100 places has meant that they have been carefully protected by the exclusion of contaminating noise sources. While this clearly does nothing for areas outside the lucky locations of these sounds (including, for example, wave sounds at Kojigahama Beach, soughing of the wind through pine trees by the ruins of Oka Castle, and singing frogs and wild birds of the Hirose River), one should bear in mind that Japanese attitudes to what and where noises are appropriate differ from those in the West: while formal gardens and teahouses are regarded and protected as the abodes of peace, many bars, cafés, and restaurants are accepted as extremely noisy places, and buses are routinely fitted with external loudspeakers broadcasting their approach with loud bursts of music.[15]

Other than noise that affects them directly, the area in which people are most concerned about noise today is one that would probably astonish Babbage, Rice, and Connell, since it does not affect humans at all: the impact of underwater noise on marine creatures.

Under the surface

Humans have been adding to the sounds of the seas for centuries, and for centuries too this has had profound effects on marine life, but it is only very recently that it has attracted much publicity. This is in part because of a genuine change: a century ago impacts were significant but always local, and were confined to very few areas, but now many of the sounds made by humans can travel enormous distances, and, while the noises of individual ships are usually quite quiet, the vast numbers of them make shipping lanes extremely noisy places. Additionally, the vast importance of sound to the lives of undersea creatures has only recently been appreciated. As a result of this new awareness, an environmental impact assessment is now usually required before activities that may generate underwater noise may commence.

This heartening level of interest about the impact of noise on marine life was highlighted on 20 January 2006, when a whale made headlines the world over when it swam up the Thames to London. A rescue was attempted but the animal eventually died from dehydration, muscle damage, and kidney failure. Though the trigger for its journey was and is uncertain, widely aired at the time was the view that it was the disorienting effect of artificial noise in the English Channel that was to blame. The fact that people jumped to the noise conclusion though there was no evidence to substantiate it demonstrates how widespread the awareness of this issue had become.

It is by no means only the sounds of shipping that ring through the oceans of the world: today, underwater acoustics performs most of the jobs that are carried out by electromagnetic radiation above the surface, including detection, measurement, tracking, target identification, mapping, communication, speed measurement, remote control, positioning, depth measurement, navigation, and measurements of ocean currents. This is despite the fact that many applications are severely limited either by the relatively long wavelengths of sound waves compared to electromagnetic ones (data transfer rates are therefore much lower) and the much lower speed (hence sonar images take far longer to produce than radar ones of the same level of detail). The reason for the comprehensive use of sound is simple: radio signals are very rapidly attenuated in sea water and light and other electromagnetic radiations fare little better, so sound must be used in default of anything else.

This range of applications not only means that vast volumes of the sea and all frequencies of sound are affected; it also makes it difficult to predict levels, or to identify individual sources. The multitude of different interests involved exacerbates this: only a few decades ago, most noise sources at sea that were not engines were military ones, but now powerful acoustic systems are widely used by fishermen, oceanographers, the renewable energies and offshore industries, and leisure craft too.

As a result of all this activity, in locations such as the English Channel artificial noise easily outweighs natural noise at many frequencies. In particular, noise from commercial shipping now dominates underwater ambient noise at frequencies below 300 Hz in many coastal areas, and it is just these locations that are rich in marine life.

There is no doubt that some parts of the sea are getting noisier, but the actual amount of increase is rarely known, and the question as to whether the entire ocean is getting significantly noisier is at present unanswerable, so many and so changeable are the natural and artificial sources involved. In terms of *types* of noise source, the picture is much clearer: the amount of commercial shipping, the extent of oil/gas exploration (which involves explosions and drilling), the frequency and range of military exercises (more explosions and low-frequency sonar), the amount of pile driving (especially for offshore wind farms), and the number of leisure vessels are all increasing and are likely to continue to do so.

Recently, it has been found that animals such as octopus, squid, turtles, and cuttlefish can be injured by underwater noise—when exposed to it in controlled experiments, they all suffered nerve lesions. Most concern, however, is with regard to the effect of noise on mammals.

It has long been known that intense noise can damage marine mammals physically, but, just in as in humans, the effects of quieter sounds are subtle and complex. For instance, if startled animals rise quickly, dissolved nitrogen in the bloodstream comes out of solution as bubbles, which can cause intense pain, damage, and death if not promptly dealt with by increasing the pressure again—in other words, the bends. Some stranded whales and dolphins have been found with their livers, kidneys, and other organs riddled with holes, most likely the effect of such rapid rising (see Fig. 41).[16] Recent studies have shown that in some cases naval sonar has been responsible.[17] Other detrimental effects include the disruption of the food chain and interference with communication links between mother and child. Breeding may also be affected. To understand all these effects requires a great deal of research, which is not easy to carry out. Though sonar data can be used to gauge the numbers of marine mammals, species are not always easy to identify, and there are many factors other than noise that might cause changes or movements in the areas occupied by them. Nor do the barriers between sciences help: there is often a suspicion of physicists by biologists interested in whales and dolphins, and this and other factors make obtaining data about the effect of noise on behaviour in the wild very difficult. Hence, much of the data on reactions to noise that are available are based on the responses of a fairly small number of tame animals, who have lived for many years in aquaria—and whose hearing has very probably been affected by their unusual lifestyles.

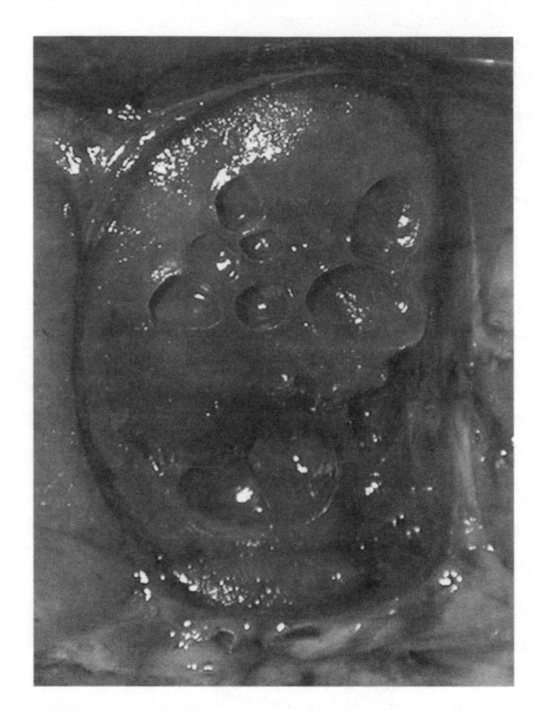

FIGURE 41. A sectioned kidney of an adult male common dolphin (*D. delphis*) stranded on a beach in Cornwall, England in 2002. It is likely that the abnormal cavities were caused by nitrogen bubbles that formed because of rapid surfacing, and are likely to have been the immediate cause of the dolphin's death. Rob Deaville and Paul Jepson, the UK Cetacean Strandings Investigation Programme (a programme funded by Defra and the Devolved Administrations).

One particular change that has had significant effects on marine life is the sudden return to the use of active sonar by military submarines. While this technology was popular in the early twentieth century, it was then largely replaced by passive sonar. However, over the last few decades submarines have become quieter, making it impossible simply to detect their engines until they are very close. As a result, active sonar was again brought to bear on the problem. Since it is the more distant submarines that are effectively silent, today's active sonars are of much lower frequencies that their earlier versions, so that they cover greater distances. This development causes particular problems for whales: animals far away are affected without the people responsible for the noise even being aware of their distress, and the low attenuation of the signals makes them inescapable. Yet, since so little is understood about the specific uses of and sensitivities to underwater sounds by marine life, the scale of this problem was not appreciated until very recently.

For instance, it was only in 2011 that new research showed that Navy sonar systems mimic sounds produced by predatory killer whales: it has been known for some time that beaked whales have stranded during naval sonar exercises, but no causal association had been established. However, by listening for the distinctive clicks of the whales (using hydrophones on the sea floor), it was possible to monitor the animals while sonar exercises were going on. Some whales were also exposed to recorded versions of the sonar scans. Others were tagged with equipment that measured both the sound levels they were exposed to and their movements and orientations. The results were clear and consistent: the whales stopped foraging for food, moved away, and rose towards the surface.

There are still great holes in the data available about the interactions between marine mammals and underwater noise But, despite this, the negative impact of artificial sound is an issue that the public and scientists alike are taking seriously, both to protect the ecosystem in order to persevere fishing stocks and because they are concerned about the sufferings of dolphins and whales. One might speculate why the impact of noise on these marine species is of more concern than, say, the great and frequent distress suffered by sheep, horses, and cattle when flown over by military aircraft. Recent studies of the impacts of such over-flying on birds and wild animals show that they too are significantly affected: ovenbirds and squirrels, to take two recent examples, decline sharply in breeding and foraging activities respectively during periods of intense over-flying. Perhaps it is the fact that marine animals are not felt to inhabit our world in the same way and that their own realm is more unspoilt—and hence that our encroachment on it is more of an invasion.

Expressions of concern regarding noise effects on divers are also notable by their rarity. Quite why this is so is not very clear. Perhaps people don't think of divers as innocent victims like dolphins and whales. What is more likely, though, is that people assume that appropriate hearing protection is provided for divers as a matter of course, just as it is for pneumatic-drill operators. This is very far from being the case, however. The actual assessment of the effect of noise on divers is plagued by many of the same sorts of problem that face those studying other marine species, and divers' hearing is often damaged by the many pressure changes their ears suffer, so that it is often hard to distinguish these effects from noise-related ones. However, in some cases the hand-held power tools used by divers at work can certainly produce dangerously high sound levels—and there are no legal thresholds to limit such noises and relatively little research into methods of doing so.

A problem with units

A problem that plagues discussions of noise levels underwater even more so than their airborne equivalents is the unavoidable complexity of the units used. Enormous noise levels have been quoted in the press, and many erroneous comparisons made. The result is that many very real problems become obscured, exaggerated, or unnecessarily contentious.

The fundamental problem is this: decibels are not things in themselves like watts; they are measures of a difference—a difference between the thing you want to talk about and some reference level, and they are meaningless unless it is clear just what that reference is. For airborne sound, the reference level (0 dB) is roughly the threshold of human hearing, which happens to be around 20 micropascals. The threshold of human hearing underwater is not of much interest, so it is not used as the starting point for the scale of underwater sound levels. Instead, the 0 dB point there is defined as 1 micropascal.

But, of course, this means that the same sound pressure that acousticians label 0 dB in air would, on the face of it, be labelled as 20 dB under water (though the next few paragraphs show that the case is more complex and the number is different). So, it is essential that sound levels stated in decibels include a mention of the reference pressure. But they seldom do. Some acousticians would say at this point that this is exactly why watts should be used instead.

A second factor is that water is vastly more dense and more incompressible than air, which means that sound travels through it in a different way. This increases still further the difference between decibels in water and decibels in air.

Another very practical issue that often causes errors in comparing noisy things, and not just when one is in air and the other is in water, is that it is essential to know how distant the sources are. In underwater acoustics, the concept of source level is a very useful one—but where is one to measure it? It would be as impossible as it would be pointless to measure the sound at the centre of a submarine or a whale, so the usual approach is to refer to the source level as the level that would be measured at a distance of 1 metre from the source. But again this is often not stated, simply because it is always used—in water. In air there is no such standard distance, and it all depends on the context. It would make little sense to talk about the noise of a jet plane 1 metre away. So, the actual distance must be stated. But often it isn't. In fact, many posters and tables that give typical value of noise sources leave out this crucial information.

Impacts of such misunderstandings can be very significant. For instance, in 2007, referring to an experimental sonar source that produces very loud low-frequency sound, the UK *Economist* wrote: 'It has a maximum output of 230 decibels, compared with 100 decibels for a jumbo jet',[18] the implication being that the sound was three *million* times 'louder'—which would mean any nearby whales would be affected just as they would be by having three million jet engines next to them. In fact, the sound was really more like thirty-two times louder, for a nearby whale—clearly still very significant, but a useful basis for setting up suitable protections and modifications, rather than an unbelievably nightmarish scenario that could be resolved only by an outright ban on such systems. And this has almost happened in other cases:

> Low Frequency Activated Sonar (LFA) is a system used by the US Navy to detect submarines. Yet it is one with which until recently others were unfamiliar. Hence, when the Navy announced that a trial was to be carried out at sea using it, a hornets' nest was stirred up, and the Natural Resources Defense Council (NRDC) took the Navy to court and succeeded in getting LFA banned. A 1999 article in *USA Today* claimed that, 'After noise pollution from shipping, LFA is of greatest concern among marine scientists.'[19]

But the actual problem was that the NRDC, the marine biologists, and the media all thought that LFA was the same thing as mid-frequency sonar—which really *is* very loud. LFA, on the other hand, is apparently not of concern to passing whales and has even been confused with natural underwater sounds by trained marine biologists.

While this may well have been an innocent error, what happened next surely could not have been, as the lawyers working for the NRDC submitted all sorts of disturbing and apparently damning evidence about the beaching of beaked whales. What was not mentioned, though, was that this has taken place in the Bahamas. The LFA ship was sited off the coast of California.[20]

It is particularly unfortunate that the facts and figures of underwater noise are often misunderstood or misused like this, considering that, compared to their airborne equivalents, there are so few of them. As a result of this scarcity, the real efforts to mitigate the impacts of underwater noise that are frequently made by those who are responsible for generating it are rendered very challenging by the large number of unknowns involved. The fundamental approach is, unsurprisingly, zoning. Before a significant new noise source—for example, a piling operation—gets the go-ahead, attempts are made first to predict the sound field over the surrounding area. This is the easy bit. The next stage is to determine the likely populations of different species present and to work out the area in which the predicted noise levels will be sufficient to cause adverse effects on them. This depends on the knowledge available about the animals' thresholds at different frequencies, which is very hard to establish, especially for effects such as startle. However, once zones of impact have been established (and these will be different for different species), then mitigations are made within those areas.

One of the simplest of these mitigations is the 'slow start', which simply means that, whatever the interfering noise source is, it is introduced gradually to the ocean environment. This can be very simple—gradually turning up the power of a sonar system, for example. The idea is that this both reduces startle and allows animals time to escape, once the noise has bothered them but before it has caused damage or significant distress.

The other major mitigation is the employment of marine monitoring officers (MMOs), whose job it is to scan the surrounding sea for animal life and halt the production of noise if an animal is seen. But, of course, this is limited to near-surface animals and is very challenging in poor visibility.

In addition, either passive or active sonar systems can be used to assess the local marine populations—the problem with these is that some creatures are simply too silent for passive sonar to be effective, while other creatures may well be disturbed by the active sonar itself.

And all of this is plagued by the lack of knowledge as to what noise—especially low level noise—actually does to marine animals.

Only yet more research—conducted by physicists and biologists working together—can supply such information.

New technologies of noise

It is not only noise produced unintentionally that can be a menace: acoustic weapons that were until recently confined to the pages of science fiction are now fast becoming reality.

In 2005, for instance, the captain of a ship off the Horn of Africa used an extremely powerful, highly directional acoustic system to discourage Somalian pirates. The Long Range Acoustic Device, or LRAD, was developed for the US military and first used by it in 2003. It has now been installed in hundreds of military, commercial, and private ships. LRAD generates a focused sound beam, effective at up to almost 9 kilometres and intensely loud—and highly unpleasant, to say the least—over shorter distances, and it can be used to project either sounds or commands. The LRAD was developed in response to a 2000 attack on the USS *Cole*, off Yemen.

A more frightening device was reported in 2008: called MEDUSA (Mob Excess Deterrent Using Silent Audio), it is a gun that fires short microwave pulses into living things, rapidly compressing and releasing tissue and causing acoustic pulses in the skull that are heard as sounds. The inventor claims that normal audio safety limits do not apply, since the sound does not enter through the eardrums—and the sound cannot be blocked as it doesn't come from outside. It was offered to the military and to farmers, for scaring birds.

In 2010, another bird-scaring device, called the Thunder Generator, was developed in Israel by agricultural technology firm PDT Agro. The Israeli Ministry of Defence then approved military versions to be constructed for use against people (see Fig. 42). The Generator works by detonating a mixture of air and liquefied petroleum gas in specially shaped 'impulse chambers', producing up to 100 intensely loud shock bursts per minute.[21]

FIGURE 42. The Thunder Generator.

Thankfully, not all new sound technology is weaponry: in 2007, University of Utah physicists developed small devices that turned sound into electricity. Waste heat from radar systems, laptops, or cooling towers is used to make sound in a heat engine and a piezo electric crystal turns it into electricity. Then in 2010, the first 'acoustic laser' was developed by a team led by Tony Kent and his colleagues at the University of Nottingham. The idea is to amplify one single sound frequency. The 'saser' is based on a thin, layered lattice made of two semiconductors, gallium arsenide and aluminium arsenide, and it can generate a sound in the terahertz region for a few nanoseconds. Like the laser before it, at this early stage in its development, the uses of the saser are none too clear; saser researcher Jérôme Faist of the Swiss Federal Institute of Technology in Zurich commented: 'They will find applications, but honestly

I don't know where or for what.'[22] However, two possibilities are enhanced ultrasonic imaging and accelerated electronic systems.

Meanwhile it is both cheering and rather surprising to discover that even now, after centuries of study and decades of science, new ways to tackle noise directly are still being developed. One such solution, which could potentially reduce noise from both traffic and industrial facilities, is the sonic crystal. Though sonic crystals sound very 'New Age', they actually work well and are in principle very simple. They consist of a periodic array of cylinders, which attenuates sound at wavelengths dependent on their spacing. The arrays work through a combination of multiple scattering and absorption by the cylinder material, rather like an anechoic chamber's wall. Furthermore, the cylinders have an attractive, sculptural quality (it was a sculpture that was the original source of the discovery of the effect). The wind can also blow through them, they can be seen through, and they can be made from recycled materials. One design consists of an array of metal tubes, punched with holes and filled with rubber crumbs. Sound crystals have also been used to build an instrument that 'recycles' noise to improve soundscapes, absorbing ambient sounds, modifying them to enhance their more pleasant elements, and then re-radiating them into the environment. The instrument, called the 'Organ of Corti', won the Noise Abatement Society's 2011 Innovation Award.

Work continues too to develop quieter technologies, from motorbikes to lawnmowers—and even alarms. Brigade Electronics (UK) has developed a range of vehicle alarms that emit the 'shushing' sound of white noise: much less annoying than tones, but still well able to alert the unwary. Used primarily on reversing alarms to date, they may well be deployed on electric vehicles in future.

Acoustic levitation, in which an intense ultrasound beam applies sufficient radiation pressure to lift small objects, has been possible for many years—but it has had few applications until recently. Now, it can be used in the International Space Station, where the microgravity conditions provide an environment in which the very small, but highly controllable forces involved are sufficient to allow small objects to be lifted without touching them. In 2010 it was also suggested by researchers at the University of Vermont as a method of keeping surfaces of robots and landers on Mars dust-free.

So, in the early twenty-first century, noise is even more wide-ranging in its uses and impacts than ever before. It is a topic that has become both more complex and more relevant as time has passed. We now have vast experience and understanding of its many aspects and the technology to control it and to make it serve us. But mastering noise is not simply a matter of technology: to keep it in its place, many things, from public discussion to integrated management and from surveys to software engineering, are required. With so many factors and so many vested interests on many sides of the noise debate, its future is not easy to predict—but not impossible.

17

A QUIETER TOMORROW?

It seems safe to say that, as a tool and as a component of music, noise has a future of ever more precise application and general usefulness. After thousands of years of development in music, and several decades of the use of noise as a delicate probe and a powerful yet precise destroyer, both these areas are on the whole well understood. The same is, of course, not true at all with those forms of noise that we wish to be rid of. Here, despite millennia of awareness, centuries of attempted control, and decades of determined scientific study and legislative grappling, some such noise is almost as much of a problem as ever. Meanwhile, other similar problems, such as stench (a very significant urban problem until a century or so ago), radioactive contamination, and the chlorofluorocarbons that punctured the ozone layer in the twentieth century, are much better controlled and, for most people, simply not a problem anymore. Why? Why is it that noise, of all the pollutants, is still such a problem today that hardly any of us are free from it?

There are many answers. Where loud noises are concerned, some of our reactions are hard-wired: ears evolved largely as a warning system, so noise is intrinsically disturbing. What the smell of burning is to the nose, noise is to the ears. This makes it quite impossible to ignore, and perhaps gives dissonance in music its unsettling qualities too. Maybe this is why the knee-jerk response to a noise problem is simply to move it away—to zone it.

Another fundamental problem lies in the way our hearing mechanism works: its extraordinary sensitivity means that we can detect the slightest flow of wave energy in the air or water, energies that are so exceedingly modest that, even if just the tiniest fraction of the power of a plane, train, or windblown tree is converted to such flows, we can clearly hear them.

The non-linear response of our hearing mechanisms adds another layer to this problem. The inevitable consequence of this is that, while removing 90 per cent of litter, CFCs, or burglars is 90 per cent of a good job done, removing 90 per cent of an annoying sound is barely noticeable. Hence, for a noise-blighted person, house, or community, only a complete answer is adequate: even a perfectly sound-proofed building is hardly better than a shoddy one if the smallest of its windows is broken. Similarly, the peace of the quietest village road can be ruined by a single ill-fitting air-conditioner, hair-trigger car alarm, noisy stereo, or budding recorder-player.

Furthermore, the loudness of noise cannot be expressed by any simple unit. The commonest way to express audible sound is the decibel and, though the A-weighted version goes some way towards assigning a number that relates to subjective loudness, it is a very rough and ready one. While the level of noise can be measured by sound level meters, the dependence of our reactions on our preconceptions and the irremovable impact of its context mean an autonomous determination of the loudness is well beyond our resources. While there are special corrections for the effects of hearing a noise indoors, or of its source being a train or plane, these can be added only by a skilled investigator who has analysed the sound field and decided which components are produced by such sources. Nor are the corrections involved negligible, often amounting to well over 10 dB, which is more than the effect of many noise mitigations.

The role of noise in society also makes it difficult to tackle. Noise is a source of power for the noise-maker. Unlike a poster, it cannot be ignored. All those groups in society who either have power or strive for it use noise: motorists, campaigners, football crowds, crowd-controllers... And, from storm-gods to church bells and rough music, from trumpets to gramophones and megaphones, this is how it has always been. So noise is made deliberately by many, whether Watt miners who liked their pumps screechy, the owners of ghetto-blasters and boom cars, or those who shout into their mobiles in restaurants. The flipside of the power of noise is that those who complain about it are regarded as weak. And the same is true of campaigners—apart from the obvious fact that they can hardly patrol the streets like politicians at election time, bellowing 'Make less noise' through their megaphones, they are also automatically regarded by many as old-fashioned and futile.

Being quiet is a most self-effacing way of being nice, very different from clearing up a local river or nursing a sick animal. The greatest reward of a quiet person is to be no bother—which hardly differs from going unnoticed. Even professionals involved in noise reduction usually have to look to their own measurements and knowledge as a source of satisfaction, of a job well done. The opening of the M6 toll road in the UK in 2003, for example, meant that the number of people impacted by high levels of noise fell sharply, since fewer cars now used the M6 and the toll road passed further away from built-up areas. But the change is said actually to have caused many more complaints—because, while a few people were now suffering more, and hence complained, the large majority who were suffering slightly less, unsurprisingly, did not write in with letters of thanks.

Complaints are, in fact, a very poor guide to noise. Neighbourhood noise is by far the commonest cause of complaint in the UK, the USA, and many other countries, yet actually the results of surveys make it clear that transport noise usually has a far greater impact. Presumably the causes of this high complaint level are twofold. Neighbourhood noise is often difficult and/or embarrassing for the sufferer to deal with directly—weighed down as he is by the still prevalent assumption that he is a nuisance for complaining at all and that what he is doing is asking a favour, not asserting a right. On the other hand, that would-be (or would-not-be) complainant is well aware that his own particular problem *could* be dealt with quite easily by the authorities. Meanwhile, the authorities know that, if they do deal with one such complaint, they should—and would be asked to—deal with all similar ones. This is an area so tricky that, with rare exceptions, governments and local authorities have fought shy of dealing with all but the most serious cases, leaving the neighbours themselves to get on and sort things out, while putting obstacles in their way, such as the regulations in force in many countries that require the inclusion of details of neighbour disputes and noise problems in the documentation required when a house is to be sold.

As a result, in some places neighbourhood noise is the main untackled noise issue. Historically, in most countries, traffic noise and then workplace noise were the first and second to be dealt with—and neighbourhood noise last or not at all. Yet this is one of the few types of noise where there is clear evidence that it is becoming more of a problem. In 1948 in Britain, 40 per cent heard noise from neighbours and 19 per cent were disturbed by it.[1] By 1991, 81 per cent of respondents heard such noise from neighbours and/or other people nearby and 37 per cent were bothered, annoyed, or disturbed to some extent.[2] A similar survey made in 1999/2000 showed that the proportion of respondents who reported being adversely affected by noise from neighbours had increased, while for all other categories of environmental noise the proportion had not changed.[3]

The time and place that noise is made are usually highly regulated by social codes—so no noise may be made by the audience during an artistic performance, but it is expected as soon as it has ended. Belching loudly after eating may generate approval, amusement, or disgust depending on the cultural norms of one's companions. In principle, then, some forms of noise could become very much reduced were the social codes to be modified to make them unacceptable. Far-fetched? Maybe, but so, a century ago, would have been the ideas that it would one day be considered unacceptable to smoke in restaurants, make racist comments, or leave dog excrement in the gutter. However, such a change will not happen simply through education or a new commandment. Nor will the plea to 'be nice' have much impact. To imagine that it would is to ignore the fact that people who listen to punk rock or brass bands are not (usually) trying to be annoying—they think what they are listening to already *is* nice, and not noise at all. This, of course, also applies to the use of dissonance in music—Stravinsky was not trying to annoy his audience with *The Rite of Spring*; he really hoped that they would like it (or perhaps a more accurate description of his feelings, and those of the punk/brass/opera enthusiast today, is that he thought they *should*). We do not and never can all agree about noise. We differ fundamentally in our opinions about all sorts of things, of course, from haircuts to horoscopes, but in those cases we can agree to differ. But when playing the music our neighbours hate or enjoying our new wind-chimes, we cannot help but inflict our choices on others. In short, one person's right to express himself and to do what he wants in his own home cannot coexist with his neighbours' right to peace. And, yet, they are both still rights nevertheless.

None of this is to say that it is pointless to promulgate the message that people should be more considerate to and aware of each other, but it is not enough on its own. It often seems to have been resorted to in the past in default of anything better, even though it could not have worked for racist jokers, smokers, speeding bikers, or non-recyclers. People would have to think and feel that they are in the wrong, and in the minority too, just as they do now in these cases. This would not only discourage them from making noise but encourage others to tell them off, too—just as the reserved British will speak up against smokers, but only if a non-smoking rule is clearly in force.

What has been done with global warming is a useful comparison here. From the situation just a few years ago when hardly anyone knew about it and fewer still were worried, now everyone has an opinion and most are in broad agreement. It has galvanized people from schoolchildren to UN officials as never before—and all this despite the fact that many who feel strongly about it are not currently directly affected by it at all.

Not only is it not possible to agree in all cases on what sorts of 'noise' we do not want; it can be equally hard to decide what sorts, and level, we do. With the exception of light, this is not true of any other pollutant: we would be entirely happy to live in a world free of litter, but we would hardly feel at home in a silent one. Do not many of us quite like concerts or powerful cars or fast planes or the occasional overwhelming dawn chorus? And do we not need phones, and alarms, and hooters? What, in fact, do we want? This is a topic on which many have something to say—scientists, psychologists, engineers, planners, physicians, politicians, environmentalists—not to mention the public at large. And yet at the same time it is clear that, even if all these groups got together and discussed the problem constructively, the questions to which they would dearly love to find answers in fact have none. What is the acceptable level of noise? What is the ideal soundscape?

In the past, a key way to deal with environmental noise problems was through the application of scientific principles and studies. Sound-reduction techniques, along with measurement technology and surveying, were often able to find new ways forward. Scientists could then work with authorities to limit noise and discourage its makers. But, such has been the hardening of society's attitude to scientific and other authority since the 1960s that this is no longer the case. Whether this is regarded as a shift in the power base from experts to public, a new cynicism, or a new freedom, it makes it far more difficult to deal with noise problems. Our changing society also means that zoning, the solution for centuries to so many forms of noise problem, is less successful now: it is easy to control night-time noise in a traditional society where people work from nine to five and live in the suburbs, where pubs close at 10.30 p.m., and where everyone is asleep by midnight; it is much less so when people set their own schedules. If we cannot answer questions like 'where do people live?' 'when do they commute?' and 'when do they sleep?', then the timing and placement of quiet zones become much more challenging.

Is noise pollution, then, an insoluble problem? Looking back over the history of noise reveals just how far we have come: workplaces, appliances, cars, planes, and concerts are all now subject to strict noise limits. Houses are better built, vehicles are quieter, noise effects are understood, the effectiveness of mitigation measures clear. There are many laws and regulations in place with clear instructions and strict penalties. But, of course, as Marie Curie wrote in 1894, in a letter to her brother: 'One never notices what has been done; one can only see what remains to be done.'[4]

Internationally there is increasing uniformity between approaches, especially within the European Union (EU). But a number of major inconsistencies do remain. For instance, in the UK, however diligent the efforts of noise consultants, architects, and builders to keep new premises quiet and adhere to the spirit and letter of regulations, and however much local people are consulted, the law still permits anyone to take the premises' owners to court later and to get that building pulled down again. In many other countries, all the builders must do is to prove that they followed all appropriate regulations, in which case they are safe from prosecution. Nevertheless, in general, the ways in which individual countries deal with noise problems are becoming increasingly similar, as the significance of the problems and the effectiveness of certain solutions become more evident.

The general EU approach of seeking or enforcing international agreement on strategic noise issues and large-scale, long-term programmes, with lower-level decisions being taken locally, is, despite numerous shortcomings and backwards steps, a truly progressive change that one can only hope other regions will adopt. Where the EU has failed—at least in some member states—is to push the knowledge and impact of these problems down to a grass-roots level. Dealing with noise should be very much a 'big society' project. The commitment of local authorities is as variable as it is crucial, so the more noise becomes more a local election issue the better.

There are many ways in which individuals may be induced to care more about noise. In the past, the number of those concerned has grown with the identification of each new group affected. The sick, children, marine animals—those who care about them care too about noise. As research identifies further specially affected groups, the ranks of those who care will grow. Or perhaps we need a loud spokesman—as Al Gore was for global warming—to stimulate debate. Or a charismatic figure, like John Connell or Julia Rice, for people to follow.

One factor that hampers attempts to control noise and reduces the seriousness with which it is taken is that, with the exception of the usual suspects such as factories and airports, there is little generally available information about what noise levels actually are. While the EU-mandated noise maps of towns and cities are publically available, they are mostly of value in identifying and controlling high-level transport noise. They are not designed to take into account other noise sources in general, such as factories or neighbours (and it is not easy to see how they could so do, since they are based on predictions). What would be of great benefit would be noise maps based on actual measurements—ideally maps that show both noise levels pertaining right now, and what the ranges, limits, and averages of the levels were over recent days or years. However, since such measurements would need to be made at many places, and from permanently installed monitoring units, the cost of using conventional instruments (sound level meters and data-loggers) to do this would be prohibitive. But technology is moving rapidly on, and a different measurement solution may be on the horizon. MicroElectroMechanical Systems, or MEMS, are microscopic structures built by the same techniques as are used for making electronic chips—which means that they can be constructed very cheaply and reliably and in vast numbers. They have already revolutionized many application areas, from hard disk drives to camera motors, and MEMS microphones are to be found in every one of the five billion or so mobile phones in the world. One idea, promulgated by the UK National Physical Laboratory, is to use MEMS microphones in place of the pricey and delicate measurement microphones currently employed for noise measurement, so that, instead of an expensive sound level meter being wielded by an even more expensive acoustician, compact, cheap, robust, radio-inked devices could be scattered wherever noise data are required and the results automatically uploaded to the Internet. They could even capture energy from sunlight and would be much less liable to being stolen or vandalized than existing devices—and would be so cheap and numerous it would hardly matter if they were. If such devices were used, measurements could be made in a vast range of situations, from monitoring the actual noise to which a home is subject, to providing real-time noise maps of cities in great detail.

In a sense, one might argue that the whole approach of noise mapping is rather strange, when the whole *raison d'être* of the European approach was originally to reduce the number of people exposed to high levels of noise. So what needs to be known to assess the problem and the effectiveness of efforts to tackle it is the amount of noise that individuals experience, not the amount of noise that places are exposed to. As nearly everyone spends each day in several different noise environments, a personal dose-meter would capture the data that really matter—just as radiation badges monitor radiation exposure. Such devices would be tricky to design and would ideally have to be very close to, or on, the ears to measure the correct sound level, but would be an excellent source of data. These too could be MEMS devices.

Just as important as measuring and reducing the noise we are exposed to is preserving what peace we have. A recent survey found that only 1 per cent of respondents had no quiet areas nearby, while 57 per cent had a quiet garden, 38 per cent could access quiet countryside nearby, and 32 per cent could visit a local park.[5] Such resources are vital, not just because a relief from noise can be a great restorative, but because, in noise issues, it is control that is key. Unwanted noise is bad; inescapable noise is terrible.

Thinking needs to be more coordinated, not just in terms of different pollutants and concerns, but with respect to different authorities too. Governments tend to pass responsibility for final decisions about, for example, wind farms to local authorities, which will naturally downplay the national concern in favour of local ones. This can turn everyone into a NIMBY, and can also waste a great deal of time and resources. In some cases, modest noise predictions lead to the refusal of planning permission, while in other cases higher noise levels are not considered relevant. This approach also means that many conflicting precedents are being set, with the consequence that there is little chance for sets of acceptable specifications to evolve naturally. Progress is possible only if governments grasp the nettle and decide on quantitative criteria, following full but finite discussion with local authorities and communities. If a clear set of such criteria were drawn up and applied nationally, planners and engineers would know from the outset how to design an acceptable wind farm for a site—or that such a design was impossible.

One aspect of a coordinated approach to noise along with other pollutants and issues is the likelihood of the extension of the soundscape concept to include other factors too—such as the attractiveness, safety, and accessibility of an area. The already well-developed idea of manipulating a soundscape in order to make noise more tolerable by adding pleasant sound elements to it is, in any case, likely to be adopted more widely as a relatively cheap and effective solution in cases where noise reduction is very challenging. A key challenge in this area is to deal objectively with the multitude of complex, dynamic, and interacting elements that together compose a soundscape: until ways to define and measure them and their effects are found, it will be difficult not only to design and implement soundscapes efficiently but also to quantify their benefits and value.

Technology will also help greatly in the direct tackling of some noise sources. Road traffic noise is a good target, because, unlike airways, road traffic volumes are not set according to the amount of noise they make, so any technological improvement is likely to deliver more peace—or at least, less disturbance—to those living nearby. Electric cars are far quieter than chemically powered ones, although they cannot be very successful until solar energy capture and battery technologies are much improved; at the moment the electricity made is dirty to produce and the batteries are so inefficient that frequent topping-up is required. The sounds they make can be carefully chosen to cause as little annoyance as possible, through being directional, short range, unvarying, and tuneful. Another very effective way to reduce road traffic noise is through resurfacing roads with quiet materials. Again, the more pressure there is from local residents, the more likely this will be. Quieter tyres could also deliver major reductions in noise: a recent UK Transport and Road Laboratory report estimated that such tyres could deliver benefits ten times greater than their costs.[6]

In the area of aircraft noise, in many countries there is an effective stalemate: the compromises are already in place, the regulations written, the measurement systems in action, and nothing but a little shifting is to be expected, so implacable are the views on either side. The only prospect of breaking such stalemates is through new technology—but that will work only if it goes hand in hand with a firm government policy on flights. Otherwise halving the noise impact of planes will simply result in a doubling of their numbers.

There certainly are prospects of developing aircraft that are significantly quieter. In 2006 hope was in the air with the Silent Aircraft project, a joint research programme by Cambridge University and the Massachusetts Institute of Technology (MIT) (see Fig. 43). Not only would the noise be significantly reduced, but improvements of around 25 per cent in fuel consumption were predicted. The plan was that the silent plane would enter service in around 2030.

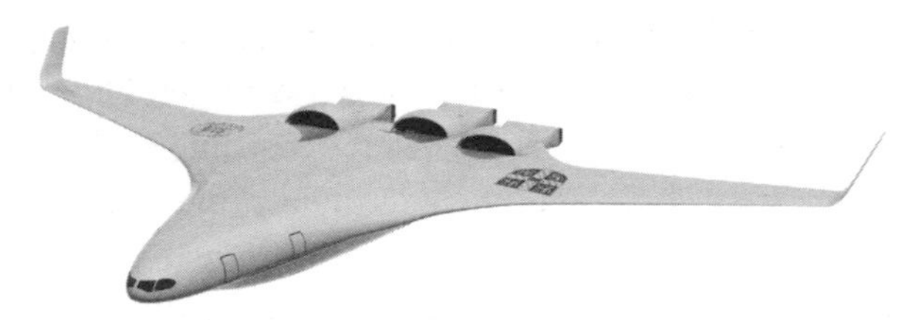

FIGURE 43. Silent concept aircraft, SAX-40 (Silent Aircraft eXperimental 40).

The aircraft's design is of the flying-wing type (experimented with but abandoned in the 1950s, as such a structure is very hard to fly without sophisticated computers). This means that the whole body provides lift, not just the wings. The absence of flaps reduces landing noise, and the engines are positioned over the wings and are of a new design, key to which are variable-sized nozzles, allowing slower jet propulsion (and therefore less noise) during take-off and climb. Though the project did not lead directly to the manufacture of a full-scale prototype, its findings have been taken up by several other organizations and may well lead to quieter skies in the years ahead.

Another major step forward in reducing aircraft noise could be the use of helium-filled airships as cargo-carriers, which would also have significant advantages in terms of reducing greenhouse gases and lowering the cost of transporting food. They would also reduce the pressure on airports, since they do not need runways to take off or land. Boeing and Lockheed Martin, supported by a large grant from the US Defense Department, are currently developing new designs for such vehicles.

On the other hand, rail noise is likely to become a greater problem in the future, with the development of high-speed rail links between many centres, in many countries. This is not simply a problem of an increase in the total amount of railway noise: at high speeds, the 'aerodynamic' noise of the train's passage through the air becomes dominant, which means that all high-speed portions of the routes of such trains are affected. With lower-speed trains, dominant noise sources are from wheel–rail interaction, engines and other onboard systems, and signalling, which often occur at more localized points, such as curves and stations.

There is reason to hope for a better future for underwater noise control, thanks to the Marine Strategy Framework Directive, which aims to achieve Good Environmental Status (GES) in Europe's seas by 2020; the concept of GES is a complex one that is still evolving in respect of noise, but it would certainly have a significant noise-control component. And simply the collation of information, the increase in monitoring, and consistency of approach by different nations are all key advantages in themselves.

However, noise control costs money. So who will pay for it? In some cases, profit will emerge as an added benefit of the noise control itself, as has happened already for many quieter vehicles and other machines. Elsewhere, though, one source of funds could be the noise-makers themselves. The principle of 'the polluter pays', already applied to many other forms of pollution worldwide, would mean that the person or organization responsible for noise pollution would be fined or taxed—and the money then ring-fenced and used to improve the situation of the sufferer(s). In France this is done already, with major airports fining aircraft that exceed agreed noise levels. The main requirement for such an approach to become more widespread is the establishment of more extensive dose–effect relationships for noise and, from them, more accurate valuation of the financial cost and health impacts. Only then could the appropriate charges be applied to noise-makers.

We now have all the technology and most of the will we need to solve the problems of noise, so that, one day, it may exist only as our tool. We really can have a quieter tomorrow—if we can work together to build it.

NOTES

Introduction

1. G. W. C. Kaye, 'The Measurement of Noise', *Proceedings of the Royal Institution of Great Britain*, 26 (1931), 435–88.
2. T. J. Cox, 'Bad Vibes: An Investigation into the Worst Sounds in the World', *Proceedings of the 19th International Congress on Acoustics, Madrid* (Physiological and Psychological Acoustics, PPA-09-003; Madrid, Sociedad Espanola de Acustica, 2007) <http://www.sea-acustica.es/WEB_ICA_07/fchrs/papers/ppa-09-003.pdf> (accessed June 2011).

Chapter 1

1. The frequency of light is around a trillion times higher than that of audible sound.
2. Some people have the ability to activate their acoustic reflexes at will—very handy when they are aware that a loud bang may be imminent. Unfortunately, this technique does not seem to be a teachable one.

Chapter 2

1. The sounds of the early Universe are of more than simply academic interest: acoustic analyses are key components of models of the evolution of cosmic structure, one of the most active areas of cosmological research today.
2. T. Fritz, S. Jentschke, N. Gosselin, D. Sammler, I. Peretz, R. Turner, A Friederici, and S. Koelsch, 'Universal Recognition of Three Basic Emotions in Music', *Current Biology*, 19/7, 14 Apr. 2009, 573–6.
3. Joyce L. Chen, Virginia B. Penhune, and Robert J. Zatorre, 'The Role of Auditory and Premotor Cortex in Sensorimotor Transformations', *Annals of The New York Academy of Sciences*, 1169, Issue: *The Neurosciences and Music III: Disorders and Plasticity* (2009), 15–34.
4. R. M. Schafer, *The Soundscape, our Sonic Environment and the Tuning of the World* (New York: Knopf, 1977), 215.
5. Robert G. Jahn et al., 'Acoustical Resonances of Assorted Ancient Structures', *Technical Report PEAR 95002* (Princeton University, Mar. 1995); see also Paul

Devereux et al., 'Acoustical Properties of Ancient Ceremonial Sites', *Journal of Scientific Exploration*, 9 (1995), 438.

6. University of Salford, Manchester, 'Measuring the Acoustics of Stonehenge', July 2011 <http://www.acoustics.salford.ac.uk/res/fazenda/acoustics-of-stonehenge/> (accessed 25 Nov. 2011).
7. *The Epic of Gilgamesh*, trans. N. I. Sandars (Harmondsworth: Penguin, 1972), 108.
8. *The Epic of Gilgamesh*, 108.
9. *nšn* ('storming/raging').
10. With thanks to James P. Allen, Wilbur Professor of Egyptology at Brown University, for this translation and for information on ancient Egyptian records.
11. One *li* was around 350 metres.
12. M. A. Stein, *Journal of the Royal Asiatic Society* (Jan. 1915), 43–4, quoted in G. N. Curzon, *Tales of Travel* (London: Hodder & Stoughton, 1923).
13. L. LaPaz, 'The Effects of Meteors upon the Earth (Including its Inhabitants, Atmosphere, and Satellites)', *Advances in Geophysics*, 4 (1958), 217–350.

Chapter 3

1. Polybius, *The Histories*, trans. W. R. Paton (6 vols; Loeb Classical Library; London: William Heinemann & Co., 1922), vol. i, bk 1, sect. 34, ll. 1–2.
2. Polybius, *The Histories*, trans. Paton, vol. i, bk 2, sect. 29, ll. 5–6.
3. Tacitus, *The Agricola and The Germania*, trans. H. Mattingly and rev. S. A. Handford (Penguin Classics; Harmondsworth: Penguin, 1970), 103.
4. Homer, *Illiad*, trans S. Butler (London: A. C. Fifield, 1898), bk 18.
5. R. M. Pidal, *Poem of the Cid*, trans. W. S. Merwin (A Mentor Classic; New York: New American Library, 1962), 16–17.
6. The history of the definition of consonance is a complex one, with at least five different meanings. Before polyphonic music developed, consonance referred to the melodiousness of sounds made successively. Afterwards, it referred to the pleasing, sonorous quality of simultaneous sounds. Then in the thirteenth century, as the rules of counterpoint were formulated, concordance referred to the clarity of a low voice in its polyphonic context. In the early eighteenth century, individual tones in a chord were referred to as consonant or dissonant, and, in the mid-nineteenth century, 'dissonance' began to mean something like 'roughness'.

7. Albert Einstein (1947); quoted from Banesh Hoffmann, *Albert Einstein Creator and Rebel* (New York: New American Library, 1972), 95.
8. Censorius Datianus, *Censorini de die natali liber*, ed. N. Sallmann (Leipzig: Teubner, 1983), xii. 1, 4.
9. Aristotle, *De Caelo*, II.9, $290^{b}16$–32, in *The Oxford Aristotle*, trans. J. L. Stocks, ed. W. D. Ross, ii (Oxford: Clarendon Press, 1930).
10. Alexander of Aphrodisias, *Greek Astronomy*, $542^{a}5$–18, trans. Sir Thomas L. Heath, (London: J. M. Dent & Sons, 1932), 34.
11. G. W. Leibniz: *Principles of Nature and Grace*, §4, in *Philosophical Papers and Letters*, ed. and trans. L. E. Loemker (Dordrecht: Reidel, 1969).
12. Plato, *The Republic*, 7.531B, in *Dialogues of Plato*, trans. B. Jowett, iii (3rd edn; New York and London: Oxford University Press, 1892).
13. Plato, *Timaeus*, 67B, in *Dialogues of Plato*, trans. Jowett, iii.
14. Aristotle, *On the Soul*, II.8, trans. R. D. Hicks, from M. R. Cohen and I. E. Drabkin (eds), *A Source Book of Greek* Science (Cambridge, MA: Harvard University Press, 1948).
15. Anaxagoras, *On Nature*, §§28–30.
16. Aeneas the Tactician (Aeneas Tacticus), *On the Defence of Fortified Positions*, XIX (Loeb Classical Library; London: William Heinemann & Co., 1923), 103.
17. Herodotus, *Histories*, bk IV, §400.
18. Vitruvius, *Ten Books on Architecture*, bk X, ch. XVI.
19. *The Architecture of Marcus Vitruvius Pollio*, trans. Joseph Gwilt (London: Priestley and Weale, 1826), bk X, ch. 16, sect. 10.
20. M. J. Crocker (ed.), *Noise Control* (New York: Van Nostrand Reinhold Co. Inc., 1984), 2.
21. Lucretius, *De Rerum Natura*, bk IV, II.
22. Horace, *Epistles*, bk II, epistle II, ll. 72–6.
23. Juvenal, *Satires*, Satire III, ll. 234–8.
24. *Law of Caesar on Municipalities*, 44 BCE, sects 14–16, in Allan Chester Johnson, Paul Robinson Coleman-Norton, Frank Card Bourne, and Clyde Pharr (eds), *Ancient Roman Statutes: A Translation with Introduction, Commentary, Glossary, and Index* (Austin: University of Texas Press, 1961), 94–5.
25. Vitruvius, *Ten Books on Architecture*, bk V, ch. III.
26. Vitruvius, *Ten Books on Architecture*, bk I, ch. I.
27. Vitruvius, *Ten Books on Architecture*, bk V, ch. V.

Chapter 4

1. Quoted from M. Riordan and Jon Turney (eds), *A Quark for Mister Mark: 101 Poems about Science* (London: Faber and Faber, 2000), 65.
2. Peter Bailey, 'Breaking the Sound Barrier', in Mark M. Smith (ed.), *Hearing History, A Reader* (Athens, GA: University of Georgia Press, 2004), 28.
3. John Stow, *Survey of London* (London: Printed for A. Churchill, J. Knapton, R. Knaplock, J. Walthoe, E. Horne, B. Tooke, D. Midwinter, B. Cowse, R. Robinson, and T. Ward, 1598), 62.
4. John Taylor, *The World Runnes on Wheeles: or Oddes, betwixt Carts and Coaches* (London: Printed by E. A[llde] for Henry Gosson, 1623).
5. D. R. Woolf, 'Hearing Renaissance England', in Smith (ed.) *Hearing History: A Reader*, 123.
6. The current interpretation of the double negative is a fairly recent innovation in English—perhaps as recent as the early eighteenth century.
7. *The Lawes of the Market Corporation of London* (London, 1595), Rule 30, ref sig. A7r–v; Rule 25, ref. sig. A6v.
8. A view held ever since: Herbert Spencer, co-inventor of the basic idea of the theory of evolution, claimed that 'you might gauge a man's intellectual capacity by the degree of his intolerance of unnecessary noises' (A. H. Davis, *Noise* (London: Watts & Co., 1937), 9).
9. Paul Hentzner, *Travels in England during the Reign of Queen Elizabeth*, trans. Horace Walpole (London: Edward Jeffery, 1797), 64–5; *The Miscellaneous Writings of John Evelyn*, ed. William Upcott (London: Colburn, 1825), 94.
10. Alain Corbin, 'Identity, Bells and the Nineteenth-Century French Village', in Smith (ed.), *Hearing History: A Reader*, 185.
11. The same measure was adopted by the richest citizens in ancient Rome. It is an effective one—but only if the gap is sufficiently wide. Sounds with wavelengths that are exact multiples of the gap between the walls can actually become much louder as a result; there is a similar problem with some secondary ceilings. This was also a problem with early attempts to use air-filled double glazing to exclude sounds. With modern double glazing, the low internal pressure acts as an effective barrier.
12. Ben Jonson, *Epicœne, or the Silent Woman*, Act I, scene I (London: William Stansby, 1616), 185.

13. T. Dekker, *The Seven Deadly Sinnes of London* (London: printed by E.A. for Nathaniel Butter, 1606), 57.
14. There are several variant spellings including chiavari, shivaree, and chivaree.
15. *Wiltshire Quarter Sessions Great Rolls*, Trinity, 1618, number 168, 'Deposition of Thomas Mills, cutler, and his wife Agnes'.

Chapter 5

1. Francis Bacon, *Natural History*, II, §114.
2. Francis Bacon, *The New Atlantis* (1627), 19.
3. Francis Bacon, *Sylva Sylvarum; Or, A Natural History, in Ten Centuries . . . Whereunto is Added . . . the New Atlantis* (9th edn; London: Printed by J. R. for William Lee, 1670), 240.
4. Francis Bacon, *Sylva Sylvarum; Or, A Natural History*, 240.
5. Johannes Kepler, *Harmonices Mundi* (Linz: Printed at the expense of Gottried Tampach, bookseller of Frankfurt, by Johannes Planck, 1619.
6. Galileo Galilei, *Dialogo sopra i due massimi sistemi del mondo* (Florence: Battista Landini, 1632).
7. Galileo Galilei, *Discorsi e Dimostrazioni Matematiche, intorno a due nuove scienze* (Leiden: Lodewijk, 1638).
8. Galileo Galilei, *Dialogues Concerning Two New Sciences by Galileo Galilei*, trans. Henry Crew (New York : Macmillan Co., 1914), 99.
9. Galilei, *Dialogues Concerning Two New Sciences*, 103.
10. Athanasius Kircher, *Phonurgia Nova, sive Conjugium Mechanico-Physicum Artis & Natvrae Paranympha Phonosophia Concinnatum* (Campidonae: R. Dreherr, 1673).
11. In fact, nature had long beaten Kircher to it: the male mole cricket digs a burrow in the form of a flared double horn, allowing him to broadcast a call at about 3 Hz. The sound exceeds 90 dB 1 metre away and can be heard at distances of at least 600 metres. The cricket also fine tunes the sound to a resonant frequency for maximum range by shaping a section of the horn's throat.
12. Sir Samuel Morland, *Tuba Stentoro-Phonica, an Instrument of Excellent Use, as well as at Sea as at Land Invented: Invented in the Year 1670 and Humbly Presented to the Kings Most Excellent Majesty Charles II, in the Year 1671* (London: Printed by W. Godbid, 1672).
13. The name derives from Stento, a famously loud ancient Greek herald.
14. Morland, *Tuba Stentoro-Phonica.*

15. Isaac Newton, *Philosophiæ Naturalis Principia Mathematica* (London, 1686).
16. Isaac Newton, *Philosophiæ Naturalis Principia Mathematica*, trans. Andrew Motte (New York: Daniel Adee, 1846), bk II, sect. VIII, proposition XLVIII, theorem XXXVIII.
17. Presumably more than its usual meaning of 'vulgarity'.
18. Newton, *Philosophiæ Naturalis Principia Mathematica*, trans. Motte, scholium to sect. VIII.
19. Letter to Henry Oldenburg, 18 Nov. 1676, in H. W. Turnbull (ed.), *The Correspondence of Isaac Newton, 1676–1687* (New York: Cambridge University Press (Published for the Royal Society), 1960), ii. 182.
20. Similarly, he put off for years the announcement of his law of universal gravitation—one of the most important individual contributions to science—because he could not predict the known speed of the Moon with satisfactory accuracy. Only when a new value of the size of the Earth became available was he willing to publish.
21. L. Euler, Dissertatio Physica de Sono (Basle: E & E. J. Thurnisiorum, 1727), trans. R. B. Lindsay in Lindsay (ed.), *Acoustics: Historical & Philosophical Development* (Stroudsburg, PA: Dowden, Hutchinson & Ross, 1973), 104–18.
22. P. M. Laplace, *Annales de Chimie et de Physique*, III (1816), trans. R. B. Lindsay in Lindsay (ed.), *Acoustics*, 181–2.
23. Robert Hooke, A *Curious Dissertation concerning the Causes of the Power & Effects of Musick*, in *The Posthumous Works of Robert Hooke, Containing his Cutlerian Lectures, and Other Discourses, Read at the Meetings of the Illustrious Royal Society* (London: Samuel Smith & Benjamin Walford, 1705).
24. Robert Hooke, *The Posthumous Works of Robert Hooke*.

Chapter 6

1. D. C. Miller, *Anecdotal History of the Science of Sound* (New York, MacMillan, 1935), 37.
2. Jean-Antoine Nollet, *Leçons de physique expérimentale* (6 vols; Paris: Les Frères Guérin, 1743), i.
3. R. M. Schafer, *The Soundscape, our Sonic Environment and the Tuning of the World* (New York: Knopf, 1977), 190.
4. A toy in the form of a horse's head on a stick.
5. Ben Jonson, *Bartholomew Fair, A Comedy*, Act III, sc. vi.
6. Jonathan Swift, *Journal to Stella*, ed. Herbert Williams (1766; 2 vols; Oxford: Clarendon Press, 1948), 581.

7. Hugh De Quehen (ed.), *Samuel Butler: Prose Observations* (Oxford: Oxford University Press, 1979), 127.
8. Barnardino Ramazzini, *De Morbis Artificum Diatriba* [Diseases of Workers], trans. and ed. Wilmer Cave Wright (1713; Chicago: University of Chicago Press, 1940).
9. W. P. Williams, *Reports of Cases Argued and Determined in the High Court of Chancery* (3 vols; London, 1740–9), i. 390, cited in E. Cockayne, *Hubbub: Filth, Noise and Stench in England, 1600–1770* (New Haven and London: Yale University Press, 2007), 113.
10. Robert Monson, Edmund Plowden, Christopher Wray, and John Manwood, *A Briefe Declaration for What manner of Speciall Nusance concerning private dwelling Houses, a man may have his remedy by Assize, or other Actions as the Case requires* (London: Printed by Tho. Cotes, for William Cooke, 1639).
11. 'A Gentleman of the Temple', *Public nusance, considered under the general Heads of Bad Pavements, Butchers infesting the Streets, the Inconvenience to the Public occasioned by the present Method of billeting the Foot Guards, and the Insolence of Household Servants, with some Hints towards a Remedy and Amendment* (London: Withers, 1754).
12. Quoted in Raymond A. Mohl, *The Making of Urban America* (2nd rev. edn; Lanham, MD: Rowman and Littlefield, 1997), 121.
13. M. J. Crocker, 'Introduction', in M. J. Crocker (ed.), *Noise Control* (New York: Van Nostrand Reinhold Co. Inc., 1984), 11.
14. Georg Christoph Lichtenberg, *Lichtenberg's Visits to England*, quoted in Robert B. Shoemaker, *The London Mob: Violence and Disorder in Eighteenth-Century England* (London: Hambledon Continuum, 2004), xi. 11.
15. Shoemaker, *The London Mob*.
16. Quoted in A. Gyrowetz, *Autobiography* (1848; Vienna: Alfred Einstein, 1915; trans. and ed. Renee Anna Illa (Ph.D. dissertation, Kent State University)).
17. Cited in J. Cuthbert Hadden, *Haydn* (London: J. M Dent & Co., 1902), 46.
18. In M. R. James, 'A Neighbour's Landmark', *Eton Chronicle*, 17 Mar. 1924, the ghost simply *is* a noise: 'There thrilled into my right ear and pierced my head a note of incredible sharpness, like the shriek of a bat, only ten times intensified—the kind of thing that makes one wonder if something has not given way in one's brain. I held my breath, and covered my ear, and shivered.'
19. A. Radcliffe, *The Mysteries of Udolpho* (1794; World's Classics Edition; Oxford: Oxford University Press, 1980), 438.

20. Tobias Smollett, *The Expedition of Humphry Clinker*, ed. Lewis M. Knapp, rev. Paul-Gabriel Boucé (World's Classics; Oxford: Oxford University Press, 2009), 119.

21. *An Essay on the Construction and Building of Chimneys. Including an Enquiry Into the Common Causes of their Smoaking [sic] and the Most Effectual Remedies by Robert Clavering* (London: Printed for I. Taylor, 1779).

Chapter 7

1. Letter to Boulton, 1777, from Birmingham Central Library, The Boulton and Watt Archive and the Matthew Boulton Papers, Part 12: Boulton & Watt Correspondence and Papers (MS 3147/3/179).

2. C. H. Parry, *Collections from the Unpublished Medical Writings of the Late C. H. Parry* (London: Underwoods, 1825), 554.

3. Letter from Beethoven to Franz Gerhard Wegeler, 29 June 1801, from Emily Anerson (ed. and trans.), *The Letters of Beethoven* (New York: Norton & Co Inc., 1961), vi. 57–62, letter 51.

4. R. T. H. L. Laennec, *Traité de la auscultation mediate et les malaides des pommons et du Coeur* (Paris: Brosson & Chaudé, 1819), 284.

5. The fact that the speed of sound increases with (the square root of) elasticity explains why the speed of sound in rubber, unlike most other solids, is far lower than in water or air—only about 62 metres per second, compared to around 330 in air and 1500 in water.

6. William Hyde Wollaston, 'On Sounds Inaudible by Certain Ears', *Philosophical Transactions of the Royal Society of London*, 110 (1820), 306–14.

7. Henry Matthews, *Observations on Sound: Shewing the Causes of its Indistinctness in Churches, Chapels, Halls of Justice, &c, with a System for their Construction : also . . . Out-buildings, Storehouses, Barns, Sheds, &c* (London: Printed for Sherwood, Gilbert, and Piper, 1826), 2.

8. Matthews, *Observations on Sound*, 8–9.

9. John F. Layson, *George Stephenson: The Locomotive and the Railway* (Newcastle upon Tyne: Tyne Publishing, 1881).

10. Samuel Smiles, *The Life of George Stephenson* (London: John Murray, 1857), 34.

11. Smiles, *The Life of George Stephenson*, 93.

12. 'Competition of Locomotive Carriages on the Liverpool and Manchester Railway', *Mechanics Magazine, Museum, Register, Journal, and Gazette*, 324, Sat., 24 Oct. 1829 (cont. from previous number), 1.
13. R. M. Reeve, *The Industrial Revolution 1750–1850* (London: University of London Press, 1971), 214–15.
14. John Fosbroke, 'Pathology and Treatment of Deafness', *Lancet*, 1, 4 Aug. 1830, 645.
15. Charles Dickens, *A Paper Mill*, from *Uncollected Writings from Household Words* (Bloomington and London: Indiana University Press, 1968), 140.
16. Dickens, *A Paper Mill*, 140.
17. J. H. Girdner, 'To Abate the Plague of City Noises', *North American Review*, 165 (1897), 463.
18. R. M. Schafer, *The Soundscape, our Sonic Environment and the Tuning of the World* (New York: Knopf, 1977).
19. Walter Hancock, *Narrative of Twelve Years' Experiments (1824–1836) Demonstrative of the Practicability and Advantage of Employing Steam-Carriages on Common Roads* (London: J. Weale, 1838), 100.
20. Committee on the Problem of Noise, *Noise Final Report* (London: HMSO, 1963), 58.
21. Edwin Chadwick, *Report to Her Majesty's Principal Secretary of State for the Home Department, from the Poor Law Commissioners on an Inquiry into the Sanitary Condition of the Labouring Population of Great Britain with Appendices* (London: Printed by W. Clowes and Sons, for HMSO, 1842), 347.
22. It was in a few decades to be one key to the fundamental discovery of cosmology: the light of clusters of galaxies is reddened to an extent that is greater for more distant clusters of galaxies, showing that all of them must be taking part in an expansion of the Universe.
23. 'The Dictionary of Victorian London' <http://www.victorianlondon.org/population/population.htm> (accessed 28 Nov. 2011).
24. Letter to Alexander Carlyle, 14 Dec. 1824, in 'The Carlyle Letters Online', <http://carlyleletters.dukejournals.org/cgi/content/long/3/1/lt-18241214-TC-AC-01> (accessed Feb. 2011).
25. Letter to Margaret A. Carlyle, 9 Nov. 1843, in 'The Carlyle Letters Online', <http://carlyleletters.dukejournals.org/cgi/content/full/17/1/lt-18431109-TC-MAC-01> (accessed Jan. 2011).

26. Gustav Theodor Fechner, *Elements of Psychophysics* (2 vols; Leipzig: Breitkopf und Härtel, 1860).
27. In 1888 Max Carl Wien completed this work by measuring Fechner's constant for different tones and found that it was not really a constant, tending to decrease as frequency increased and also increasing at very low and also very high intensities.
28. *American Historical Documents, 1000–1904, with Introductions, Notes and Illustrations* (Washington: Library of Congress, 1910), xliii, para. 96.
29. Charles D. Ross, 'Sight, Sound and Tactics in the American Civil War', in Mark M. Smith (ed.) *Hearing History: A Reader* (Athens, GA: University of Georgia Press, 2004), 267–78.
30. Which the eye does not: we see the same yellow colour irrespective of whether the light entering our eyes is at the 'yellow' wavelength or, if no such wavelength is present at all, is replaced with a mixture of red and green light (of the appropriate hues and brightnesses).
31. Aristotle, rather impressively, was aware of this distinction, even though there was no way for him to make—or hear—a pure tone: 'Why does the low note contain the sound of the high note? It is like an obtuse angle, whereas the high note is like an acute angle'; and, even more specifically: 'Why is it that in the octave the concord of the upper note exists in the lower but not vice versa?'
32. Jane Austen, *Mansfield Park* (1st edn; London: T. Egerton, 1814), vol. III, ch. VIII, p. 176.
33. Such matters of etiquette can be turned on their heads, however. In 2011, Alexander Lukashenka, dictatorial leader of Belarus, made it a crime to applaud him. In a country where demonstrations or criticism of the regime are forbidden, activists had come up with the brilliant scheme of clapping in such an ecstatically over-the-top manner that it was obvious (to all but Lukashenka, for a while) that sarcasm was their intention.
34. F. Nightingale, *Notes on Nursing: What It Is, and What It Is Not* (London: Harrison & Sons, 1859), 57.
35. M. J. Crocker, 'Introduction', in M. J. Crocker (ed.), *Noise Control* (New York: Van Nostrand Reinhold Co. Inc., 1984).
36. US Bureau of Census Online: 1840: <http://www.census.gov/population/www/documentation/twps0027/tab07.txt>; 1860: <http://www.census.gov/population/www/documentation/twps0027/tab09.txt > (accessed 3 Apr. 2012).

37. Florence W. Asher, *Women, Wealth and Power: New York City, 1860–1900* (New York: City University of New York, 2006), 19.

Chapter 8

1. J. H. Clapham, *Economic History of Modern Britain*, ii. *Free Trade and Steel 1850–1886* (Cambridge: Cambridge University Press, 1932).
2. Charles Manby Smith, *Curiosities of London Life, Physiological and Social of the Great Metropolis* (London: Cash, 1853), Music Grinders §5.
3. Charles Babbage, *Passages from the Life of a Philosopher; Street Nuisances*, in *The Works of Charles Babbage*, ii. *Passages from the Life of a Philosopher*, ed. Martin Campbell-Kelly (London: William Pickering, 1864).
4. Babbage, *Passages from the Life of a Philosopher.*
5. Babbage, *Passages from the Life of a Philosopher.*
6. August Strindberg, *The Red Room*, trans. Elizabeth Sprigge (Everyman Edition; London: J. M. Dent & Son, 1967).
7. Benjamin Ward Richardson, *Hygeia: A City of Health* (London: Macmillan, 1876), 19.
8. John William Strutt, Baron Rayleigh, *Theory of Sound* (2 vols; London: Macmillan, 1877, 1896).
9. This also makes life tricky for the kakapo, a flightless bird from New Zealand. As its mates can be so scattered, it has to make a very low-pitched booming sound in order to be heard by them (low-pitched sounds are less attenuated by air than higher ones, which is why distant pop concerts sound so bassy). But, what physics gives to the kakapo, it also takes away—the pitch is so low that the females cannot establish its direction. So the kakapo adds some high-pitched squawks too, in the hope that the booming has resulted in some females wandering accidentally within earshot. The kakapo is almost extinct.
10. The reason why acoustic improvers added to concert halls like the Albert Hall are curved is to ensure that the listeners hear the first reflections of a sound as soon as possible after the original sound, and the round shapes, by scattering the incident waves over the whole audience, ensure that this happens as quickly and evenly as possible. In general, if the direct and first reflected sounds are heard within about 50 milliseconds of each other, the reflections reinforce the direct sound, while, after about 50 milliseconds, the sounds are heard separately, as an echo that interferes with the direct sound.

11. T. Barr, 'Enquiry into the Effects of Loud Sounds upon the Hearing of Boilermakers and Others who Work amid Noisy Conditions', *Proceedings of the Royal Philosophical Society of Glasgow*, 17 (1886), 223–39.
12. Barr, 'Enquiry into the Effects of Loud Sounds', 224.
13. Barr, 'Enquiry into the Effects of Loud Sounds', 224.
14. Barr, 'Enquiry into the Effects of Loud Sounds', 230.
15. R. M. Ballantyne, *Blown to Bits: Or the Lonely Man of Rakata, A Tale of the Malay Archipelago* (London: James Nisbet & Co., 1894), 140.
16. Karl Bücher, *Arbeit und Rhythmus* (Leipzig: S. Hirzel, 1896).
17. Anon., 'The Horseless Carriage and Public Health', *Scientific American*, 18 Feb. 1899, 98.
18. Raymond W. Smilor, 'Cacophony at Thirty-Fourth and Sixth: The Noise Problem in America, 1900–1930', *American Studies*, 18 (1977), 23–38.
19. Thomas Hawksley, *Catalogue of Otoacoustical Instruments to Aid the Deaf* (3rd edn; London: John Bale, [c.1895]), 3.
20. Such experiments must have been flawed, as few adults can hear above 15 kHz. Children can often hear up to, but very rarely beyond, 20 kHz. It is likely that the tuning forks generated lower-pitched subharmonics along with 30 kHz ones.
21. Francis Galton, *Inquiries into Human Faculty and its Development* (London: Macmillan, 1883), 49.
22. These are typical figures; there is a dependence, too, on the volume of the concert hall. For instance, organ music sounds best with a reverberation time of about 1.6 seconds in a 1,000-cubic-metre space—but, for a space ten times the volume, the ideal time is around 2.0 seconds.

Chapter 9

1. A diffraction effect, thanks to the reducing density causing a slowing-down of the light—rather like a car whose left-hand wheels are retarded and that hence turns to the left.
2. Or Audion tube, as De Forest called the first version, in reference to its use in audio circuits. This caused rather a fuss from the classically inclined, who complained that it was half-Greek and half-Latin, and suggested, in the person of a Dr Michael I. Pupin, a fellow electronic engineer, 'If he had said acouion or acousticon it might have been better', to which Lee De Forest responded: 'When we use a term one hundred times a day, it is necessary to have something brief.'

3. Richard Birkefeld and Martina Jung, *Die Stadt, der Lärm und das Licht* (Seelze: Kallmeyer, 1994), 45, cited in K. Bijsterveld, *Mechanical Sound* (Cambridge, MA: MIT Press, 2008), 98.
4. Edward A. Abbott, 'The Yelling Peril', *American City*, 6 (Mar. 1912), 575.
5. Anonymous, 'Noise', *American City*, 2 (June 1910), 41.
6. G. Trobridge, 'The Murder of Sleep', *Westminster* Review, 154/3 (1900), 298.
7. Trobridge, 'The Murder of Sleep', 300.
8. Karl Bücher, *Arbeit und Rhythmus* (Leipzig: S. Hirzel, 1896).
9. *Le Figaro*, Paris, 20 Feb. 1909.
10. Arseny N. Avraamov, 'The Emerging Musical Science and a New Era of Music History', *Muzykal'ny Sovremennik*, 6 (St Petersburg, 1916).
11. Anonymous, 'Noise and Factory Efficiency', *Scientific American Supplement*, 76/20 (Sept. 1913), 189.
12. A. Garcia (ed.), *Environmental Urban Noise* (Boston: WIT Press, 2001), 10.
13. In 1911, Proust livened up his dull evenings by subscribing to a 'theatrophone' service, which provided a telephonic link to eight Parisian theatres and concert halls (William C. Carter, *Marcel Proust: A Life* (New Haven: Yale University Press, 2000)).
14. Michael Haberlandt, *Cultur im Alltag: Gesammelte Aufsätze von Michael Haberlandt* (Vienna: Verlag, 1900), 177–83.
15. C. Masterman (ed.), *The Heart of Empire: Problems of Modern City Life in England* (London: T. Fisher & Unwin, 1901), 137.
16. N. W. McLachlan, *Noise: A Comprehensive Survey from Every Point of View* (London: Oxford University Press, 1935).
17. Every three seconds is about 1 kilometre.
18. Imogen B. Oakley, 'Public Health vs the Noise Nuisance', *National Municipal Review*, 4 (Apr. 1915), 231–7.
19. BBC News, Latin America and Caribbean, 'Buenos Aires is "Noisiest City" in region, Study Says', 29 Dec. 2010 <http://www.bbc.co.uk/news/world-latin-america-12087598> (accessed 15 Nov. 2011).
20. Dan McKenzie, *The City of Din: A Tirade against Noise* (London: Adlard and Son, Bartholomew Press, 1910).
21. Roger Kamien, *Music: An Appreciation* (abridged edn; New York: McGraw-Hill, 1989), 250.

22. Luigi Russolo, *L'arte dei rumori* (Milan: Edizioni Futurista di 'Poesia', 1916).
23. Quite apart from the tiny power of the voice, one also has to take account of the fact that just a small fraction of the total power output is at the resonant frequency of the glass—even if that is the fundamental (lowest) pitch of the sung note, most vocal energy is spread through the harmonics.

Chapter 10

1. Robert J. Urick, *Principles of Underwater Sound* (3rd edn; New York: McGraw-Hill, 1983).
2. Gregory Haines, *Sound Underwater* (London: Scientific Book Club, 1975), 36.
3. The others being: *Allied Submarine Devices Investigation Committee, Anti-Submarine Detection Indicator, Anti-Submarine Detection Investigation Committee,* and *Allied Submarine Detection Investigation Committee.*
4. Haines, *Sound Underwater,* 40.
5. W. Plowden (1971), *The Motor Car and Politics 1896 – 1970* (London: The Bodley Head, 1971), app. 1.

Chapter 11

1. P. Bagwell and P. Lyth, *Transport in Britain 1750–2000* (London and New York: Hambledon Continuum, 2002), 162, citing Air Ministry Reports.
2. Michael J. T. Smith, *Aircraft Noise* (Cambridge: Cambridge University Press, 2004), 21.
3. Hugo Fastla, Sonoko Kuwanob, and Seiichiro Nambac, 'Railway Bonus and Aircraft Malus for Different Directions of the Sound Source?', paper at the Exposition and Congress on Noise Control Engineering, Rio de Janeiro Brazil, 7–10 Aug. 2005.
4. Department for Environment, Food and Rural Affairs, *Noise and Nuisance Research Newsletter* 235: *Research into the Improvement of the Management of Helicopter Noise* (London: HMSO, 2008) http://archive.defra.gov.uk/environment/quality/noise/research/documents/nanr235-project-report.pdf (accessed June 2011).
5. L. Daston and P. Galison, 'The Image of Objectivity', *Representations,* 10/40 (1992), 81–128.
6. There was considerable debate as to what type of logarithm would best fit animal reactions. In the end, two different units were defined, the decibel, which used logarithms to base 10, and the neper, which used Naperian

logarithms (that is, logarithms to base e). In fact, as the relationship is not exactly logarithmic, either unit is equally useable in practice, and the Neper faded into the obscurity where it resides today.

7. E. J. Richards, 'Noise Annoyance and its Assessment', in A. W. Haslett and Johen St John (eds), *Science Survey* 2 (London: Vista Books, Longacre Press Ltd, 1961), 274.
8. Emily Ann Thompson, *The Soundscape of Modernity: Architectural Acoustics and the Culture of Listening in America, 1900–1933* (Cambridge, MA: MIT Press, 2002), 146.
9. D. A. Laird and E. L. Smith, 'Acoustical Stimuli which Affect Stomach Contractions', *Journal of the Acoustical Society of America*, 1 (1930), 256.
10. Philip Bagwell and Peter Lyth, *Transport in Britain 1750–2000* (London and New York: Hambledon Continuum, 2002), 99.
11. Committee on the Problem of Noise, *Noise Final Report* (London: HMSO, 1963), 168.
12. Robert Musil, *The Man without Qualities*, trans. S. Williams and B. Pike (1930; rev. edn, London: Pan Macmillan, 1997).
13. Lord Horder, 'Noise and Health', *Quiet*, 1/5 (July 1937), 11.
14. Unfortunately, this contribution was to be short lived; the Noise Abatement Commission (formed in 1929) was dissolved in the same year—one more victim of the depression.
15. Through the amendment of the Road Traffic Act in 1934, which required exhausts to be fitted with mufflers, banned noisy vehicles, and prohibited the use of the motor horn in built-up areas between 23.30 and 07.00.
16. E. F. Brown et al. (eds), *City Noise: The Report of the Commission Appointed by Dr Shirley W. Wynee, Commissioner of Health, to Study Noise in New York City and to Develop Means of Abating It* (New York: Noise Abatement Commission, Department of Health, 1930), 80.
17. P. R. Bassett and S. J. Zand, 'Noise Reduction in Cabin Airplanes', *Transactions of the American Society of Mechanical Engineers*, 56 (1934), 49.
18. Bassett and Zand, 'Noise Reduction in Cabin Airplanes', 49.
19. Brown et al. (eds), *City Noise*, 74.
20. Frederic Charles Bartlett, *The Problem of Noise* (Cambridge: Cambridge University Press, 1934).
21. N. W. McLachlan, *Noise: A Comprehensive Survey from Every Point of View* (London: Oxford University Press, 1935), 134.

22. Baron William Thomson Kelvin, from 'Electrical Units of Measurement', a lecture delivered at the Institution of Civil Engineers, London (3 May 1883), quoted in American Association for the Advancement of Science, *Science* (Jan.–June 1892), xix. 127.
23. Though such units still inevitably simplify what is really a very complex relation between perception and physical phenomena: for instance, though pitch does depend strongly on frequency, it is also affected by the intensity of the sound. A low-frequency tone will decrease in perceived pitch as its intensity increases, while the pitch of a high-frequency tone increases.
24. In 1988, soon after I started working in acoustics, I attended a meeting to discuss how to make the meaning of the decibel clearer for government officials to understand. A senior acoustician, who had been noticeably quiet throughout the meeting and was well known for his correct and cautious manner, suddenly bellowed out to everyone present: 'This is exactly the same sort of discussion I've been having for the last twenty years! If people *still* don't get the point, we might as well all go up and go home!'
25. Assuming, that is, that one knows the acoustical properties of the object through which the sound is to pass. When this object is one of us—a complicated melange of bone, muscle, and fat—calculating the point of focus becomes very complex, and one must use special tissue-mimicking materials to do so.
26. Anon., 'Yonsei University plans first OCD clinical trial', *Focussed Ultrasound Surgery Association Newsletter*, 33 (July 2011) <http://www.fusfoundation.org/Newsletter/fusf-newsletter-volume-33> (accessed Aug. 2011).

Chapter 12

1. *Manual of German Radio* (1937).
2. Gregory Haines, *Sound Underwater* (London: Scientific Book Club, 1975), 82.
3. Carl Eckart (ed.), *Principles and Applications of Underwater Sound* (Washington: Summary Reports Group of the Columbia University Division of War Research, 1946), 252, sect. 13.6.4, p. 252.
4. Eckart (ed.), *Principles and Applications of Underwater Sound*, 253, sect.13.6.4.
5. Sir William Bragg, *The World of Sound* (New York: G. Bell and Sons Ltd, 1920), 141.
6. Richard C. Paddock, 'Undersea Noise Test Could Risk Making Whales Deaf: Scripps Institution Researchers Propose the Experiment to Study Global Warming. Critics Voice Alarm', *Los Angeles Times*, 22 Mar. 1994, p. 1.

7. D. Chapman, *A Survey of Noise in British Homes, National Building Studies, Technical Paper Number* 2 (London: HMSO, 1948).
8. Mori, National Noise Survey, 2008 <http://www.ipsos-mori.com/researchpublications/researcharchive/2253/National-Noise-Survey-2008.aspx> (accessed June 2011).

Chapter 13

1. K. Fellinger and J. Schmidt, *Klinik und Therapies des Chromischen Gelenkreumatismus, Maudrich* (Vienna: Maudrich, 1954).
2. G. Maintz, 'Animal Experiments in the Study of the Effect of Ultra-Sonic Waves on Bone Regeneration', *Strahlentherapie*, 82 (1950), 631–8.
3. There is some evidence that ultrasound can be heard through bone conduction, but it is hard to ascertain whether what is perceived is the ultrasound itself or its subharmonics.
4. M. S. Fox, 'The Wisconsin Story of the Industrial Noise Problem', *Noise Control*, 1 (Jan. 1955), 74–6, 89.
5. Harry F. Olson and Everett G. May, 'Electronic Sound Absorber', *Journal of the Acoustical Society of America*, 25 (1953), 1130–6.

Chapter 14

1. Leo L. Beranek, *Noise Reduction* (New York, Toronto, and London: McGraw-Hill, 1960), 1.
2. Leo L. Beranek, *Music, Acoustics and Architecture* (New York, 1962), 5
3. Beranek, *Music, Acoustics and Architecture*, 1.
4. Not all the problems were due to the overall shape: others arose because a number of shapes that were to have decorated the side walls and would have acted as useful acoustic diffusers were not installed, in what turned out to be a very costly cost-saving measure.
5. The noise is produced by unsteady mixing between the jet of accelerated air and the still air around it.
6. Committee on the Problem of Noise, *Noise Final Report* (London: HMSO, 1963), 133.
7. *Thomas Arthur Down* v. *Dudley Coles Long Ltd*, Devon Assizes, 27–31 Jan. 1969.
8. See Rainer Guski, 'Community Response to Environmental Noise', in A. Garcia (ed.), *Environmental Urban Noise* (Boston: WIT Press, 2001), 111–48.

9. J. Katska et al., 'Longitudinal Study on Aircraft Noise Effects at Dusseldorf Airport 1981–1995', *15th ICA Proceedings* (Trondheim: Acoustical Society of Norway, 1995), iv. 5.
10. Guski, 'Community Response to Environmental Noise'.
11. The NAS later (1972) suggested a larger network of smaller airports with all passengers a maximum of one hour away from one, both to spread the noise burden and to reduce road congestion and overall car use. But the expense of such a plan was prohibitive. Cost also stymied the proposal that all major long-distance roads should run along concrete-lined cuttings with overhanging tops.
12. HMSO/Noise Advisory Council, *Aircraft Noise: Should the Noise and Number Index be Revised?* (London: HMSO and Noise Advisory Council, 1972), 5.
13. United States Code, Title 42, Chapter 65, §4901.
14. US Environmental Protection Agency (Washington, 1981), EPA/981-101.
15. R. M. Schafer, *The Soundscape, our Sonic Environment and the Tuning of the World* (New York: Knopf, 1977).
16. The exact speed depends on air pressure and temperature.
17. E. J. Richards, 'Noise Annoyance and its Assessment', in A. W. Haslett and John St John (eds), *Science Survey* 2 (1961), 292.
18. Richards, 'Noise Annoyance and its Assessment', 292.
19. From transcript of PBS documentary *Supersonic Dream*, US airdate 18 Jan. 2005 <http://www.pbs.org/wgbh/nova/transcripts/3203_concorde.html> (accessed 15 Nov. 2011).

Chapter 15

1. Rupert Taylor, *Noise* (Harmondsworth: Penguin, 1970), 240.
2. A. Garcia (ed.) *Environmental Urban Noise* (Boston, WIT Press, 2001), 202.
3. White noise is a mix of all frequencies, and sounds like rushing or hissing.
4. Joint Committee on Human Rights, Parliament of the United Kingdom, *Counter-Terrorism Policy and Human Rights: Terrorism Bill and Related Matters: Oral and Written Evidence* (London: The Stationery Office, 2005), 110.
5. Gregory Haines, *Sound Underwater* (London: Scientific Book Club, 1975), 36.
6. Ronald H. Cole, 'Grenada, Panama, and Haiti: Joint Operational Reform', *Joint Force Quarterly* (United States Department of Defense, Winter 1998–9), 61.

7. US SOUTHCOM Public Affairs after Action Report Supplement, 'Operation Just Cause', 20 Dec. 1989–31 Jan. 1990, pp. 4–6.
8. At least, at moderate speeds: at over about 20 or 30 knots, the propeller causes the formation of micro-bubbles and the sound produced wipes out any hope of a stealthy approach.
9. Road Traffic & Aircraft Noise & Children's Cognition & Health (RANCH) Project, European Commission 5th Framework Project: Quality of Life and Management of Living Resources—Key Action 4: Environment and Health <http://www.wolfson.qmul.ac.uk/RANCH_Project/> (accessed Mar. 2012). See, e.g., S. A. Stansfeld, B. Berglund, C. Clark, I. Lopez Barrio, P. Fischer, E. Ohrstrom, M. M. Haines, J. Head, S. Hygge, I. van Kamp, and B. Berry, 'Aircraft and Road Traffic Noise and Children's Cognition & Health: Exposure–Effect Relationships', *Lancet*, 365 (2005), 1942–9.
10. C. Clark, R. Martin, E. van Kempen, T. Alfred, J. Head, H. W. Davies, M. M. Haines, I. Lopez Barrio, M. Matheson, and S. A. Stansfeld, 'Exposure–Effect Relations between Aircraft and Road Traffic Noise Exposure at School and Reading Comprehension: The RANCH Project', *American Journal of Epidemiology*, 163/1 (2006), 27–37.
11. 'Future Noise Policy', European Commission Green Paper, COM (96) 540 final, 4 Nov. 1996.
12. Lden is a weighted twenty-four-hour average of the noise levels during the days, evenings, and nights—55 dB Lden is judged by the EU to be the level at which noise becomes annoying.
13. Following this, the Environmental Protection Act 1990 made it the responsibility of local authorities to investigate noise complaints. But investigation was one thing; reduction was quite another. Local authorities objected that the existing laws on noise were not sufficiently effective to deal with noise, especially neighbourhood noise, where a key issue was the time taken to deal with complaints, which could easily run into months. The suggestion of the government working party set up to handle the Act and its challenges was that this could be alleviated by making night-time noise-making an offence in itself, without the need to make a complaint, and this became law with the passing of a new Noise Act in 1996. However, the issue of delays is not the only one in this area: the fact that such complaints must be reported to prospective purchasers of a property often puts people off making them, as does, obviously, the likelihood of falling out with neighbours.

Chapter 16

1. A technological solution that confines the sounds to the immediate vicinity of the phone is also conceivable; it has been proven that it is theoretically possible to create an invisible 'cloak' that excludes sound waves (Duke University, Pratt School of Engineering, '"Invisibility Cloaks"' Could Break Sound Barriers', 2011 <http://www.pratt.duke.edu/node/281/?id=1934> (accessed May 2011).
2. Vic Tandy and Tony R. Lawrence, 'The Ghost in the Machine', *Journal of the Society for Psychical Research*, 62/851 (Apr. 1998), 360–4.
3. Tandy and Lawrence, 'The Ghost in the Machine'.
4. A. Tanura, 'Effects of Landscaping on the Feeling of Annoyance of a Space', in A. Schick and M. Klatte (eds), *Contributions to Psychological Acoustics. Results of the 7th Oldenburg Symposium* (Oldenburg: Bibliotheks- und Informationssystem der Universität Oldenburg, 1997), 135–61.
5. J. Kastka et al., 'Comparison of Traffic-Noise Annoyance in a German and a Swiss Town: Effects of the Cultural and Visual Aesthetic Context', in H. Höge and G. Lazarus-Mainka (eds), *Contributions to Psychological Acoustics. Results of the 4th Oldenburg Symposium* (Oldenburg: Bibliotheks- und Informationssystem der Universität Oldenburg, 1986), 312–40.
6. Directive 2008/1/EC of The European Parliament and of The Council, 15 Jan. 2008, concerning integrated pollution prevention and control <http://eur-lex.europa.eu/LexUriServ/LexUriServ.do?uri=OJ:L:2008:024:0008:0029:EN:PDF> (accessed Apr. 2011).
7. World Health Organization: Regional Office for Europe and Charlotte Hurtley, *Night Noise Guidelines for Europe* (World Health Organization, 2009) <http://www.euro.who.int/__data/assets/pdf_file/0017/43316/E92845.pdf> (accessed Feb. 2011).
8. This measure attempts to measure the impact of factors like noise that can cause ill-health, as well as or rather than early death. Though potentially very useful, values are extremely hard to establish or even define properly <http://www.euro.who.int/__data/assets/pdf_file/0008/136466/e94888.pdf> (accessed June 2011).
9. European Foundation for the Improvement of Living and Working Conditions, *Second European Survey on Working Conditions 1996* (European Foundation for the Improvement of Living and Working Conditions, 1997) <http://www.eurofound.europa.eu/pubdocs/1997/26/en/1/ef9726en.pdf> (accessed Feb. 2011).

10. Other available figures are: Germany 16%, France 14%, Belgium 12% (A. Garcia, *Environmental Urban Noise* (Boston: WIT Press, 2001), 112).
11. For instance, all member states must use the same noise indicators: Lden (day–evening–night equivalent level) and Lnight (night equivalent level).
12. F. McManus, 'Noise Law in the United Kingdom—A Very British Solution?', *Legal Studies*, 20/2 (June 2000), 264–90.
13. French Environment and Energy Management Agency, 'Noise: Facts and Figures' <http://www2.ademe.fr/servlet/KBaseShow?sort=-1&cid=96&m=3&catid=17565> (accessed 15 Nov. 2011).
14. Li Jing, 'Big Cities Urged to Curb Noise Pollution', *China Daily* online, 26 Jan. 2011 <http://www.chinadaily.com.cn/cndy/2011-01/26/content_11917110.htm> (accessed Mar. 2011).
15. See, in particular, Garret Keizer, *The Unwanted Sound of Everything We Want* (New York: Public Affairs, 2010), 147.
16. P. D. Jepson, R. Deaville, I. A. P. Patterson, A. M. Pocknell, H. M. Ross, J. R. Baker, F. E. Howie, R. J. Reid, A. Colloff, and A. A. Cunningham, 'Acute and Chronic Gas Bubble Lesions in Cetaceans Stranded in the United Kingdom', *Veterinary Pathology*, 42 (2005), 291–305.
17. A. D'Amico, R. C. Gisiner, D. R. Ketten, et al., 'Beaked Whale Strandings and Naval Exercises', *Aquatic Mammals*, 35 (2009), 452–72.
18. Anon., 'Quiet, Please. Whales Navigating', *The Economist*, 5 Mar. 1998, p. 1.
19. Anon., 'Man's Roar Ripples through Whales' World', *USA Today*, 6 July 1999, p. 8.
20. Anon., 'Man's Roar Ripples through Whales' World', *USA Today*, 6 July 1999, p. 8.
21. Barbara Opall-Rome, 'A Cannon "Stun Gun": Israeli Device Harnesses Shock Waves for Homeland Defense', *Defense News*, 11 Jan. 2010 <http://www.defensenews. com/article/20100111/DEFFEAT01/1110306/A-Cannon-Stun-Gun-> (accessed 5 Mar. 2012).
22. *Nature*, '"Sasers" Set to ~Stun', 26 Feb./ 2010 <http://www.nature.com/news/2010/100226/full/news.2010.92.html> (accessed 11 Nov. 2011).

Chapter 17

1. D. Chapman, 'A Survey of Noise in British Homes', National Building Studies, Technical Paper No. 2 (London: HMSO, 1948).

2. C. J. Grimwood, 'Effects of environmental noise on people at home', BRE Information Paper IP22/93 (1993).
3. Department of Environment Food and Rural Affairs, *The UK National Noise Attitude Survey 1999/2000.*
4. Letter to Józef Sklodowska, 18 Mar. 1894, quoted in Eve Curie, *Madame Curie,* trans. Vincent Sheean (New York: Doubleday, Doran & Co.,1937), 116.
5. National Noise Survey 2008 <http://www.environmental-protection.org.uk/assets/library/documents/National_Noise_Survey_2008.pdf> (accessed May 2011).
6. Transport Research Laboratory, *Published Project Report PPR077: Tyre/Road Noise—Assessment of the Existing and Proposed Tyre Noise Limits* (June 2006).

INDEX

Bold numbers denote illustrations